WHISPERS
FROM THE
FARM

Practical advice and fond
memories from
those who have felt a call to the
rural life

WHISPERS
FROM THE
FARM

Christopher McNinch

Advantage®

Published by Advantage, Charleston, South Carolina.
Member of Advantage Media Group.

ADVANTAGE is a registered trademark and the Advantage colophon is a trademark of Advantage Media Group, Inc.

Printed in the United States of America.

ISBN: 978-159932-273-5
LCCN: 2012936527

This publication is designed to provide accurate and authoritative information in regard to the subject matter covered. It is sold with the understanding that the publisher is not engaged in rendering legal, accounting, or other professional services. If legal advice or other expert assistance is required, the services of a competent professional person should be sought.

 Advantage Media Group is proud to be a part of the Tree Neutral® program. Tree Neutral offsets the number of trees consumed in the production and printing of this book by taking proactive steps such as planting trees in direct proportion to the number of trees used to print books. To learn more about Tree Neutral, please visit www.treeneutral.com. To learn more about Advantage's commitment to being a responsible steward of the environment, please visit www.advantagefamily.com/green

Advantage Media Group is a leading publisher of business, motivation, and self-help authors. Do you have a manuscript or book idea that you would like to have considered for publication? Please visit www.amgbook.com or call 1.866.775.1696

DEDICATION

*To my grandfather, James McNinch Sr.,
my inspiration, who, at 96 years young,
is still going strong.*

*He has a wonderful way of weaving stories from his past into
conversation, and I am convinced that his being a farmer for a
major chapter in his life is the reason he is still living today.
It is my hope that I, too, will have some wonderful stories to
share with my grandchildren about my family's own small-farm
journey. I joke with him that he can live to whatever age he
wants but I am going to beat him by at least a year. He has
set the bar pretty high. Thank you for your love, patience and
guidance.*

ACKNOWLEDGMENTS

I would like to thank my family and friends for their support during the creation of this book. I also want to thank all of the wonderful people who took the time to share their own farm stories that are woven into this book. Thank you to the folks at Advantage Media Group for their help and support in getting this project off the ground. Thank you to my editor, Bob Sheasley, for his understanding and support of this project. Finally, I would like to thank all of the farmers who work hard every day to provide what our country needs to grow and prosper.

A donation will be made to the National FFA Foundation for their work to support young leaders and agriculturalists and to Heifer International for their work in communities to end hunger and poverty and to care for the Earth.

The National FFA Organization (also known as Future Farmers of America) envisions a future in which all agricultural education students will discover their passion in life and build on that insight to chart the course for their educations, career, and personal future.

The FFA makes a positive difference in the lives of students by developing their potential for premier leadership, personal growth and career success through agricultural education. Agricultural education prepares students for successful careers and a lifetime of informed choices in the global agriculture, food, fiber and natural resources systems. Visit **www.ffa.org** for more information.

Heifer International is a nonprofit humanitarian organization dedicated to ending hunger and poverty and caring for the earth. Heifer currently provides livestock, trees, seeds and training in environmentally sound agriculture to families in more than fifty countries, including the United States. Visit **www.heifer.org** for more information.

ABOUT THE AUTHOR

Christopher McNinch has heard the whisper. McNinch, 45, a longtime financial advisor, runs a fourteen-acre farm in Upstate New York raising alpacas and chickens with his wife, Lisa, and their two teenage children. The dream that called to him was a return to the land, to the farm life his grandparents knew. *Whispers from the Farm* is his quest to help others take that step, through practical advice and by sharing the stories, wise and wonderful, of those who have the same passion.

Please visit **www.farmwhispers.com**
to contact Chris and find links to the resources
referenced throughout this book.

TABLE OF CONTENTS

Selecting high-value crops
Specialized nurseries
Agritourism
Organic animal livestock
Organic hay and feed grains
Organic vegetables and fruits
Unusual or heirloom vegetables
Grapes for winemaking
Oriental vegetables
Herbs
Dried and cut flowers
Ornamental shrubs and trees
Greenhouse or hoop house
Hydroponically grown crops

109 ★ CHAPTER 4:
YOUR FARM'S FINANCIAL PLAN

FEATURED ESSAY: *We're Off to the Fair*
by Barbara Fleishman, Belmont, N.Y.

FEATURED ESSAY: *A Day at the Racetrack with Dad*
by Mary Ann Hoffman, Wellsville, N.Y.

The funds to get started
Developing a business plan
Estimating your sales
The importance of diversification
Maintaining your edge
Production plans

Your financial plan:
  Projected income statement
  Cash flow analysis
  Source and use of funds analysis
Where to get help
Professional assistance
To market, to market…

FEATURED ESSAY: *Outhouse Blues*
by Don Nielsen, Bluff Point, N.Y.

What a marketing plan should include
Internet and social media
Networking opportunities
How much can you afford?

FEATURED ESSAY: *Turkey CPR*
by John Roemmelt, DVM, Union Springs, N.Y.

FEATURED ESSAY: *Why I Don't Like Rats*
by John Kriese, Branchport, N.Y.

Overcoming obstacles
The problem of high overhead

INTRODUCTION

WE SHARE A DREAM

"Do not grow old, no matter how
long you live. Never cease to stand like
curious children before the Great Mystery
into which we were born."

– Albert Einstein

WE SHARE A DREAM

We share a dream, you and I. I know we do, simply because you have opened this book. It's a dream of days past, or ones yet to come. The farm life is whispering to your heart, calling you back to what you once knew or urging you forward to take a brave step.

You are not alone. The multitude of Farmville fans who indulge their virtual farm fantasy on Facebook is a reflection of our yearning to return to the land, start a real farm, build a backyard coop, grow the perfect tomato, tend a beehive, or milk a goat. People want a share of the life that their parents, perhaps, or their grandparents knew, or the life that they desire for themselves. Comes a time, for some, that the virtual must become reality.

I set out to write this book so that I might share some practical lessons I've learned along my own journey in starting a small farm. And if you, too, have embarked on that journey, congratulations!

You are becoming part of the foundation on which civilizations worldwide were founded.

Agriculture today is highly diversified, representing more than two hundred crops and animal options. To ensure you meet or exceed the returns made possible through farming, it is essential that you choose the right commodity or commodities. You need to grow, raise and sell something that will match your interests, skills, abilities, and resources. The rewards are not only personal but also cultural and financial. You will be playing a role in strengthening the overall agricultural community – and the economy as a whole.

This book is meant to provide guidance in helping you to make the right selections for your small-farm enterprise. Once you have identified the best commodity, it's time to put together a game plan. Whether you are a new to farming and have no agricultural experience or you already have a farm and are considering whether to expand, this book along with the reference material provided can help guide you through some of the decisions you will need to make to get started.

But this book will bring you much more than that.

Facts and figures and worksheets only go so far. They inform and explain and guide, but they don't inspire. In this book, I want to share with you the stories of people whose hearts you will understand, because they are like you. They knew a small farm, once, somewhere, sometime. In their reminiscences, they tell of good times and hard times and people they loved or loathed – tales that will make you laugh, cry and sigh.

Between the lines, they whisper wisdom and the secrets for the good life. They tell the stories that will pass down to their children so they might come to cherish the rural roots that made them who they are. This is our heritage.

I had set out, as I conceived this book, to collect dozens of such stories through a website on which farm fanciers could share their tales. I was able to collect numerous stories that way – but not nearly as many as I'd hoped. Instead, many of the stories you'll encounter in this book came to me word-of-mouth, or by letter. High-tech modes of communication aren't always the ones that suit those who harbor the memories.

One day, frustrated about how difficult it seemed to collect the wealth of stories I'd envisioned, I was feeling more than a small dose of self-pity. Nothing seemed to be going right, and I went for a walk to clear my head. Upon my return, my wife, Lisa, called me over to look at something. I really didn't want to come into the barn, but she seemed excited by her discovery so I reluctantly went to see what she had found.

I looked up, above the main entry door. There, perched on a ledge about half an inch wide, was a barn swallow nest – the one in this photograph.

I looked at that nest for a long while and was struck by the sheer will it must have taken for the swallow to attach it, piece by piece, to the barn wall above the ledge. I think I was meant to see that nest at that moment. It changed my entire mood and helped to remind me that anything worthwhile takes time and perseverance to bring forth the intended results.

If those birds could keep at it until they succeeded, I knew that I could, too. The stories I wanted to gather were out there. It would just take time and patience to collect them. Amazing how something as simple as a bird's nest can restore your perspective.

It can be hard to find such perspective during the cruel springs we sometimes have here in the Northeast. The weather can fluctuate radically from a 75-degree sunny afternoon to a 40-degree damp and rainy day the next. It plays havoc on how I feel and can influence how productive I am during the day. I am sure I am not alone in this.

Sometimes I struggle to find something that will put some cheer back in my day. And, as I had discovered by observing the barn swallow nest, inspiration can be found in the simplest of things. Such as a seed. It's about regeneration – in nature, and in the soul.

I felt it last spring when Lisa and I started some flower and plant seeds on our sun porch. It was my first experience of actually getting my hands dirty by starting some seeds indoors. I have planted many flowers and other plants outdoors, but they were usually bought at a greenhouse and were ready to be planted. This time we decided to sprout our own.

There is something spiritual about putting garden soil and some impossibly small seeds into a little container, adding water and light, and watching magic happen. Within a few days, our sprouts started peeking from the soil, adding a tint of green to our sun porch. Way better than a Chia pet, for sure.

One thing that helps to lift my spirits is music. I came of age in the 1980s, so a lot of the music that I listen to today is from that period. I take great delight in embarrassing my kids while rocking out with my air guitar or singing in the car to whatever song catches my ear. Hey, we all sound great with the radio turned up, right? As I did my exercise routine one morning, one of the songs on my music playlist was "One Simple Thing" from a band called the Stabilizers.

I began to think about that line and smiled as I thought about our little seedlings, bathed by the sunlamp on our porch. Despite the weather, that small patch of green had the remarkable ability to provide the inspiration to make a day seem bearable.

If you have never tried it, get a few packets of seeds and start your own little indoor garden. You will experience the miracle of new growth – and that may be what is inside of you, too, waiting to blossom.

Not that farming is all joy. You'll see that clearly in the stories you encounter in this book. Rain comes when you need sunshine, and vice versa. And when you get that sunny break, that weekend beyond compare, you cannot sit back. My family knows this. Last spring we finally connected two weekend days without rain, and a morning that dawned perfectly for a picnic, but what did we do?

"Hey, let's go move some poop!" I suggested after finishing breakfast.

I suppose there are farms that stay dry and manure-free. Not ours. Muddy pastures and manure-laden barns are endemic to the vocation. We had been working for several weeks around the barn and entryways in an attempt to clean up from the long winter and give our alpacas a chance to dry out.

That's what we do – raise alpacas. Why alpacas, you might ask? Well before we get into that and get in too deep, let's continue our poop story. Alpacas can produce one to two percent of their body weight in manure each day.

When you multiply that by 39 animals – well, you get the idea. Alpaca poop makes a great fertilizer for your garden, and we have seen the direct benefits of it in improving our soil in both the garden and flower beds. We even bag and sell the dried beans for making manure tea to water plants. You can visit **www.alpacatea.com** for more information.

On this particular manure-shoveling day, we let the alpacas out on some very soggy pasture to stretch their legs and get a few bites of that first tender grass. We then proceeded to move about twenty-five very full and heavy wheelbarrows full of manure and old hay to the dung pile. A definite workout for the muscles and yes we were pooped. Pun intended.

By evening it was raining again. But as I went to bed that night and reflected on the day, I felt strangely satisfied with our day's labors. We were digging into the mountain of work that farming entails, and we had made some small degree of progress.

Not that it all went smoothly. If you are farming, you know how it goes. If you plan to farm, you'll know soon enough: Sh*t happens.

And it happens even on a busy day. You can be up to your ears in manure, yet you have to do some of the other needed duties such as getting water for the animals. It's a relatively easy chore most of the time, unless you have a dog named Daisy.

Daisy is our Great Pyrenees livestock guard dog. When she arrived as a pup on our farm, we were amazed at the things she loved to chew on. We quickly discovered that just about anything left within her mouth's reach was fair game. Here's a partial list of what has passed through her mouth, and presumably out the other end: shoes, gloves, garden hose, duct tape, fiberglass panel (ouch), plastic coffee mug, hats, plastic containers, wood, mice, rabbits, and a few birds.

We naturally tried to make sure anything that was harmful or valuable was kept out of reach, but that didn't stop her from getting creative. My wife once made her a nice doggy bed of sturdy red cloth that she filled with alpaca fiber. She made it rugged enough to take any abuse, thinking Daisy would be delighted. The next morning we found that she had enjoyed it immensely. My wife came into the house with a look of disgust, holding what was left of the shredded bed. Oh well, maybe Daisy just didn't like the color.

As Daisy has aged, most of her chewing habits have stopped, but she still savors a good opportunity to keep me humble.

As we ran the hose to the alpacas' water buckets that morning, we discovered that our new $40 heavy-duty hose was actually now three hoses. Daisy had been busy. At that point, I became even busier, going back and forth carrying one bucket at a time from the hydrant to finish the watering.

Later, amid everything else I had to do, I went to town and bought a new hose. This time I mounted it on the side of the barn, out of her reach. A few mornings later, when I turned on the spigot, I was greeted with a full blast of cold water in the face. Daisy strikes again! Apparently, a length of hose had fallen just far enough from the reel so that she could stretch herself to get in a few bites.

As the saying goes, some days you bite the bear and other days the bear bites you. Or the dog bites your new hose. Sometimes it just goes like that, and you have to laugh and keep things in perspective. Especially if you are just starting out.

HOW WE GOT STARTED

So what's our story, and why did a family of newbies start an alpaca farm? That gets sort of complicated. I guess you can blame it on a pair of socks and the video music channel VH1.

During the summer of 2005, we were living in the beautiful village of Penn Yan, N.Y., and were feeling the call of having more land, a big garden for growing our own produce, and an opportunity for our children to roam wide-open spaces. My work as a financial advisor is stressful at times, and I wanted a place to come home to that gave me a feeling of peace and contentment. I know most of us have that moment in our day when it just plain feels good to be home. At the time, I was working from an office right behind my home, and while it was convenient, it never felt that I was completely away from work. I would go out to mow my lawn or play with my kids, and my office building would seem to be beckoning me back for just a while longer. I needed some separation from this and felt that moving out into the country would be just the ticket.

At the same time, my wife was at a crossroads in her life now that our children were both in school full time. She had been an elementary school teacher for twelve years and could go back to teaching but really had a desire to start her own home-based business and still be around for the kids. After many lengthy discussions and some soul-searching, we decided to begin looking for some land to build on, and we explored businesses that could be run from home. Enter the alpaca!

In 2004 my wife was diagnosed with a circulatory problem called Raynaud's disease. We call it "ghost fingers." It causes discoloration of fingers and toes. She has to be careful during the cold weather we have in Upstate New York or when handling cold items from the grocery store freezer. Her sister wanted to get her something warm for her feet, so for Christmas that year she gave her a pair of alpaca socks. Those socks soon became my wife's favorites. We'd heard of alpacas before but didn't know much about them.

When we began to contemplate a move to a more rural location and to research stay-at-home business opportunities, we thought starting a small farm might be worth exploring. My family had roots in dairy farming, but I had never really considered it to be something that would interest me. So we started to research ways to make income from a farm, listing crops we could grow and animals we might want to raise. Cows, sheep, chickens, goats and rabbits all had potential, but we knew that at some point, in order to profit from them, we would have to send them out for slaughter. We really didn't want to go down that road initially. And as newbie farmers, we wanted to start with animals that would be easy to care for and didn't require a huge learning curve.

One night we were sitting in bed watching a VH1 show called "Bands Reunited." Essentially it tried to reunite old 1980s-era bands that had broken up. This particular episode featured the rock band "Extreme." I loved their music, so I watched with anticipation. The host, Aamer Haleem, was unable to persuade the members to reunite, but one of his visits was to the alpaca farm of bassist Pat Badger.

We watched, with great curiosity, as the crew filmed Pat and his young son with these amazing animals in the background. His farm looked like something from a storybook.

"That's what we should be raising on our new farm," I told Lisa. "If a rock star can raise alpacas, so can we."

We immersed ourselves online, and visited several local farms finding out everything we could about alpacas. The more we researched, the more excited we became. Lisa loves quilting and fiber arts, so becoming "fiber farmers" seemed like a natural fit. Now we just needed a place to call home.

We looked at several properties, but nothing seemed to match our wish list of around fifteen acres, good water supply, and the same school district our children were already in. Then one day in December 2005, our real estate agent called with some news: He had a property with fourteen acres that already had an old farmhouse and pole barn on it. The online listing looked intriguing, so we decided to go have a look. What we found was an old house that reeked of dog urine and needed to be gutted. In addition, it needed a new septic system, the land was overgrown with weeds, and the pole barn was full of garbage.

Sold!

And so it began. The McNinches' small-farm adventure was on its way. As with any adventure, especially starting a small farm, you are bound to run into obstacles that will challenge you, as if to test whether you're committed. You will ultimately face that moment of truth: Are you in it for the long haul or will you bail out when one little thing goes wrong?

I had plenty such moments of truth during the ordeal of rebuilding our old farmhouse.

I remember the day well. I had come to check on the progress of our builder and his crew as they were midway through reconstruction of the farmhouse, which was built in the 1860s. I say reconstruction, because we ended up gutting and replacing over ninety percent of the original structure. Many a day went by that I was certain that we had made one of the dumbest decisions ever by purchasing a house in such poor shape. You've seen newspaper listings for the "handyman's special" or "fixer upper." As we started the demolition, it became apparent that this was more of a "tear down and start over" project. There wasn't a level spot in the house, and the foundation had suffered greatly due to water intrusion.

I had put in a full day's work at the office and wanted to come out to the farm and see the progress over the past few days. Early on, I had made the decision that I would be the general contractor, hoping to save some money and have more direct control over the building process. For the most part I am glad I did it, but it wasn't easy. At times it was just a pain in the ass.

Upon arriving at the site, I noticed that my builder's truck was parked in an odd spot in the driveway, and when I got out, I realized that the engine was running. I figured that he must be headed out for supplies and had forgotten something in the house. I went inside, and to my surprise, the house was empty, or so I thought.

I wandered around on the first floor for a minute and decided to climb the ladder to the second floor to see if anyone was up there. We had recently taken out the stairs so the only way to access the second floor was by ladder. I thought I could hear faint voices coming from the back bedroom, so I climbed up to see what was happening.

I was greeted by the work crew sprawled out around the room drinking a few beers and smoking a few alternative cigarettes, if you know what I mean. To say I was stunned would be an understatement. They all jumped to their feet and began to pick up their mess, apologizing as they scrambled. I asked what was going on and where "Joe" was. (I've changed his name to protect the guilty.) They said they were waiting for him to come back from town with more nails.

"That's funny," I said. "His truck is sitting in the driveway." They looked at one another, realizing that they were in deep and that anything else they might say would only bury them further. I told them to pick up their mess and go home. I didn't need a bunch of drunks or stoners getting hurt on my property.

I went back downstairs to see if I could locate Joe and find out what the hell was going on. I searched and called his name, to no avail. I finally went outside to see if he had gone to the pole barn or outhouse. As I walked by his still-running truck, I glanced through the side window and did a double-take.

Joe was slumped in the front seat, passed out, with the air conditioning on full blast. I wondered how long he'd been there. Carefully I opened the door and turned the engine off. By this time, the work crew had emerged from the house, just in time to see their boss in all his drunken glory. Not sure what to do, I gently shook him and started calling his name. He finally came to, saying that he hadn't slept well the night before and had needed a nap.

"Are you drunk?" I asked him. I could smell the alcohol on his breath and knew he shouldn't be driving. I chose not to confront him at that moment and asked if one of the crew was capable of driving him home. Once they all left, I went back inside to check on the progress of the work. It was minimal.

I had to summon all of my inner strength not to fire Joe and his crew on the spot. I took a deep breath and told myself that "this too shall pass" and gave them another shot. After I confronted Joe and his crew, they did settle in and do their jobs better for a while. But a few months later, the inevitable day came when I encountered a similar drunken episode. It was too much and they were all let go.

Three weeks went by with no one at the job site, putting us behind schedule. A few anxiety attacks and some stress-ridden days were to follow, but thankfully an angel appeared in the form of a new builder named John who was able to pick up the project and see it through to completion.

That's how life can be. You may think you know exactly what's happening, and it turns out to be an illusion.

People you thought you could trust let you down – not necessarily maliciously, but because humans have weaknesses and lapses. Remember how certain childhood beliefs were shattered when realities came crashing down – ones your parents may have tried to protect you from?

Here is one man's reminiscence:

NIGHTMARE IN THE GREENHOUSE

I was around age four or five when my grandfather called me one day to follow him along to the greenhouse. I assumed we were going to pick some veggies for dinner, but this day was going to be different.

On our way, he stopped by a shed and grabbed a bulging burlap bag. I figured it was filled with some of the amazing garden vegetables that my grandparents produced from their land. As we walked, I noticed that the bag would wiggle now and then, as if something alive was inside. As we approached one of the empty greenhouses, my grandfather stopped and picked up an ax, which seemed strange: He hadn't mentioned we were going to be cutting firewood.

That's when I saw it. Out from a small hole in the sack popped the head of a chicken. Before I knew it, grandpa had the bird down on a block of wood and had chopped the head right off of it.

He opened the sack, and the headless bird flew from one end of the greenhouse to the other. I stood there in shock, unable to move or speak. I think I may have wet myself but was too scared to check.

As the bird flopped, I could hear my grandmother yelling for me to get in the house immediately. As I ran for the house, I could hear her scolding my grandfather, using words that I had never heard before. Let's just say that he got an earful from her, and from my mother as well, for having done such a stupid thing in front of me.

I had nightmares for several years about what I had seen, and to this day, some sixty-five years later, I get creeped out thinking about that flight of the headless bird.

– Tom Hoffman, Wellsville, N.Y.

Yes, all is not as it seems. But then again, sometimes our deepest fears turn out to be unfounded. And such was the case one warm summer evening as my family settled down to supper. We heard a noise.

"What the heck was that?" I asked. I got only puzzled looks in return. It was a shrill sound, and we could hear Daisy barking at something in the darkness. We scrambled to collect our shoes and a flashlight and headed outside.

The sound seemed to be emanating from one of the alpacas. At first we thought a coyote had attacked, but in the dim light we could see all of the alpacas standing at alert in the barnyard.

The flashlight revealed no critters that might be stirring up Daisy – no deer, no coyote, and no rabbit. The sound seemed almost like that shrill and unforgettable cry a rabbit makes when it's about to get eaten.

The sound grew louder as we approached the barn. It was coming from Orabella, our dominant female alpaca. She was making a sound that was sort of a cross between a scream and a horse's frantic whinny – basically, an alpaca alarm call.

Turning on the barn lights, we looked for the source of the animals' horror. They barely took notice of us when we entered their pen. They stood staring at the pasture fence. We played the flashlight along the fence line, and there we saw it: a yellow plastic shopping bag that had blown into the wires. It rustled and shook with each breath of breeze.

My wife went over and plucked the bag from the fence, rescuing the alpacas from the killer plastic bag. The alpacas, and Daisy, slowly calmed down – as did we.

So it goes in the farm life. Just like Tom, who thought the greenhouse would yield something better, or like Orabella, whose fear was really over nothing, farmers endure day by day not really knowing for certain what will come their way. Good days and bad, they're all part of it.

CHAPTER 1

IS FARMING RIGHT FOR YOU?

"The small landowners are the most
precious part of a state."

— Thomas Jefferson

IS FARMING RIGHT FOR YOU?

J ust how simple is the simple life? Dreams can begin as a pure vision only to be obscured in the dust of hard work and frustration. But that need not be the case. In any endeavor, not all moments are joyous. Some are downright dumbfounding. But if you forge ahead with a realistic sense of what you want and what to expect, you can find your fulfillment in farming.

Farming really is a lifestyle choice. For me, it was an opportunity to get out of the cerebral world of what I do for a living and experience a more physical connection with the outdoors. We have ten laying chickens that produce great-tasting eggs for eating and baking, and last year raised fifty Cornish-cross meat chickens to fill our family's and friends' freezers. Lisa loves her ever-expanding garden and greenhouse, and that has allowed our whole family the opportunity to eat better, with fresh vegetables for part of the year and canned goods to enjoy over the winter months.

In addition to her love of quilting, my wife has also become an accomplished fiber artist, making beautiful things from our alpaca fiber to sell on our farm website.

My children are still very connected to the electronic world and love their iPhones, but a part of them is now grounded in basic animal husbandry and the physical effort it takes to keep the daily farm operations going. Having baby alpacas called "crias" born each year has helped bridge that awkward moment when it's time to have the birds-and-the-bees talk. Our children have been present for both births and deaths, and we hope that this involvement with the farm and connection to the land stays with them long after they leave the house to start their own lives.

We had no illusion that running a farm was going to be easy, and we are still learning from our mistakes. However, some day when I look back on this period in our lives, I am certain that the lessons learned and time we shared on the farm will have been worth it.

In this chapter, we'll begin to explore the many options to consider when starting a small farm. But first, let me share with you one man's introduction to the farm life – and his awakening, as a boy, not only to the nitty-gritty of what farmers must endure, but also to the rewards of diligence and cooperation.

MY MOST UNUSUAL FISHING TRIP

When I was about eleven years old, I had heard of a great fishing spot up near the swamps outside of town. Problem was, you had to go across a certain farmer's field to get to the pond. This farmer was noted for being difficult to deal with. He could be mean. I finally summoned the courage to ask the farmer if I could fish in his pond. I had no idea what lay ahead of me on that day.

No one was at the house, so I rode my bike up to the barn and walked around looking for the farmer. I rounded the corner of the barn, and what I saw just about made my eyes pop out. There was the farmer, lying on his side on the ground, with one arm fully inside the back end of a cow lying directly in front of him. He glanced up at me.

"What do you want?" he asked gruffly. I tried to speak, but no words would come out. He asked again, and I finally was able to mumble something about fishing.

"If you want to fish," he said, "I need your help first." He said the cow had a calf stuck inside and he needed to get the tractor and some chain to pull it out. He'd let me fish if I would hold on to the calf's feet, which were now protruding from the cow, while he went for the tractor. I reluctantly said yes.

I held on to those feet for what seemed an eternity. I finally heard the tractor coming, and over the next half hour watched, in part amazement and part horror, the extraction of a baby calf from its mother. He thanked me for my help, and I headed off. And by the way, the fishing was great!

– Jim Rysewyk, Branchport, N.Y.

When considering starting a rural enterprise, it is essential to assess your personal goals, your resources and your land. Weighing those factors will help you to save time and energy while also helping to direct your research.

Farming may sound simple enough, but there are many directions in which you can take a small farm enterprise, so consider carefully what you hope to achieve. Ask yourself questions such as these:

- ★ Do I want to spend more time with family?
- ★ Am I interested in a farming enterprise to replace or supplement my current salary?
- ★ Am I interested in farming because I want a change in lifestyle or more laid-back one?

All of those questions will help you in establishing your goals for your farm based on personal values.

This is incredibly important, and it is something that many business plans do not take into consideration.

After examining your goals, there are many other factors to consider in determining whether a small farm is the right choice for you, and if so, which type of farming enterprise you should start.

WHAT IS A SMALL FARM?

First, let's define what a small farm actually is. Definitions can be misleading, depending on your source, so it is important to understand what constitutes a small farm operation and what level of income is typically generated from the size of the operation you intend to create.

In terms of tax purposes, the definition can vary somewhat based on your local tax assessment guidelines. You should make a point to check with your local tax assessment office to determine what is defined as a small farm in your area.

According the U.S. Census of Agriculture, the definition of a farm has changed in the last 100 years. Here's how:

1850-1869...... $100 in agricultural product sales
1870-1899...... $500 in sales, or 3 acres
1900-1909...... one full-time person
1910-1924...... $250 in sales, or 3 acres, or one full-time person
1925-1949...... $250 in sales, or 3 acres

1950-1958...... $150 and 3 acres, or $250 in sales
1959-1974...... $50 and 10 acres, or $250 in sales
1975-present... $1,000 in sales

Today, the U.S. Department of Agriculture defines a small farm as one with a Gross Cash Farm Income (GCFI) that is less than $250,000 per year. According to a 2010 USDA study, 91 percent of U.S. farms are classified as small. About 60 percent of those are very small, generating GCFI of less than $10,000. These very small, noncommercial farms exist independently of the farm economy, in some respects, because their operators rely heavily on off-farm income.

This is critical to understand, especially when starting out. You will most likely need some source of off-farm income to meet your family's needs for a time. In other words, don't quit your day job just yet until you have sufficient cash flow from the farm to meet your needs. It is quite possible that you have no intention of making your farm a full-time occupation. That is perfectly OK. There are many reasons people like to farm that are more important than using it as a source of primary income. Ultimately it is up to you to define what your small farm will look like and the benefits you wish to derive.

CHOOSING WHAT YOU WILL GROW AND RAISE

After making the decision to begin a small farm, another important decision is what you will grow or raise. There are certainly many options available to you; however, prior to deciding what you will produce on your farm, it is critical to consider several factors, including what you want to achieve from your farm and what and how much you can contribute to your enterprise. As with many decisions, it will boil down to time and money.

INTERESTS AND RESOURCES

Farming can provide both personal and financial rewards, but only as long as you are willing to put in the careful thought, planning, consideration and work that is necessary.

Success also depends on designing your small farm to meet your individual needs as well as your financial resources. It has often been said that it takes money to make money, and that is no less true when it comes to small farming. You must consider the amount of financial resources you are willing and able to contribute to your endeavor.

To be successful in small farming, you must recognize that you will need to be actively involved in the operation. Focus on the activities you enjoy and consider delegating those activities that you do not. For instance, if you enjoy growing things but do not enjoy the marketing aspect of selling what you grow, you might consider partnering with or hiring someone to handle that responsibility.

To determine the type of small farm that will best suit you, consider these questions:

★ How much time can I contribute toward my small farm?
★ What types of work do I prefer?
★ How much am I able and willing to invest, both in time and money, in my small farm?
★ What scale of farming will best suit my interests and needs?
★ If I raise livestock, is there a competent veterinary clinic reasonably close that can treat my type of livestock?

PHYSICAL LABOR

There is no getting around the fact that farming on any scale requires some physical labor and interaction. How much depends, of course, on the type of farming and commodity. Some options, such as livestock, may not require as much labor, but it must be done daily. Other options, such as field crops, may require rather intensive labor, more so during planting and harvest seasons.

★ Which level of interaction and labor suits you best?

★ How much are you able to do on your own?

★ Do you have friends and family who will be able to help you during the most labor intensive times of the year?

★ Do you have a capable farm hand or farm sitter if you want to go on vacation or need to be away?

★ Are you only able to contribute time during weekends and vacation periods?

★ Would you prefer to be free during the winter months?

These are all important questions that you must ask yourself in order to determine which type and level of farming will best suit you.

You should also recognize that farming is not generally restricted to just growing the crops and raising the animals. It also involves marketing in order to make a profit. In addition, you will likely need to handle other matters, such as mechanical work on farm equipment, working with people to sell your goods – a whole variety of activities.

FINANCIALS

While it can easily cost millions of dollars to set up a large-scale or commercial farm enterprise, a small farm can be established for much less of an investment. Even so, you will need some initial capital. Consider the following questions:

- ★ How much are you willing and able to invest in your small farm?
- ★ Are you willing to borrow money to purchase equipment and develop your farm?
- ★ Can you share assets with nearby neighbors?
- ★ Is it possible to contract out some of the work without the need to purchase large or expensive equipment?

When it comes to the actual investment, note that crop productions do not typically require as much in terms of fixed assets as greenhouse and livestock operations. There are even some crop productions that do not require much of an investment at all, such as a small-scale vegetable or herb garden.

SCALE

Another important consideration when planning a small farm, is the scale at which you wish to operate. Ask yourself whether you want your farm to produce your primary income or whether you merely want to supplement your existing non-farm income. We looked at our farming enterprise as an opportunity to make some supplementary income that would cover expenses and leave enough to use toward other discretionary family expenses. It has also been a way to learn and grow in our appreciation of the benefits of knowing where a portion of our food comes from.

WHAT CAN BE SOLD?

No matter what size farm you ultimately decide to operate, in the end you will need to consider what you can realistically sell from your farm to make a profit. This means analyzing the market and identifying opportunities. Regardless of how much you might like the idea of growing a particular crop or raising a certain type of livestock, you need to consider whether there is a market for it.

In most instances, a large-scale commercial farm will produce a product that is much like what other farmers in the area produce, and it will sell its products through a rather formal marketing process. A smaller farming operation might also use a formal marketing system, but many also choose to market in different ways, including selling products right from their property via a farm stand.

Direct marketing will usually allow you to gain a higher price for your product. You can achieve direct marketing through markets that have been well-established, such as roadside stands, farmers' markets, local produce stores, restaurants, etc. You might also consider trying to break into a new market niche that may not currently be served, such as growing specialty mushrooms.

SKILLS AND RESOURCES

You must also consider your own skills and the resources that are available to you and which will give you the greatest competitive advantage. Farming can involve a wide variety of activities, including working with the farm equipment, dealing directly with people, feeding the livestock, planting and harvesting crops, etc. Carefully consider your own physical abilities and skills.

In addition, you must also consider the types of products that would be best suited to production on the property you have available. Some crops, of course, grow best in certain types of soil, and some need more space. The same is true of livestock.

Consider the following when analyzing your potential small-farm property:

★ Location: Is it in a busy or quiet area?
★ What types of water resources do you have available?
★ What about drainage?
★ What is the climate like?
★ How many frost-free days do you have?

★ Do you have early warming soils in your area?

★ What is the history of production in your area?

Beyond the resources available in terms of your skills and your farm property, you must also consider the types of products and services that will be most in demand in your community. It doesn't make sense to produce a commodity on your farm that is not in demand. Both niche and direct-market opportunities exist right in your own community, but it is imperative to identify a valid product or service that you can profit from before starting your small farm.

This will require research, involving the analysis of demographics, culture, geography, activities, tourist attractions, and location. Our research turned up only a small number of alpaca farms operating in our area and at the time none were focused on delivering high quality products made from their exquisite fiber.

Following are some of the more common agricultural products and services you might consider as options for your small farm. This is not meant to be an exhaustive list, but a place to start your research. As you review the list, ask yourself whether you would be able to provide a competitive advantage in producing one or more of these commodities:

SMALL-FARM COMMODITY OPPORTUNITIES

LIVESTOCK

- ★ Dairy cows
- ★ Beef cattle
- ★ Hogs
- ★ Alpacas – fiber, breeding, boarding
- ★ Custom grown lambs
- ★ Sheep – meat or wool
- ★ Goats – meat or dairy
- ★ Eggs – traditional, organic or free range
- ★ Broiler chickens, organic or free range
- ★ Turkeys – traditional, organic or free range
- ★ Horses – breeding, boarding
- ★ Horses – training and riding lessons
- ★ Rabbits

EXOTICS

- ★ Emu
- ★ Ostrich
- ★ Llama
- ★ Bison

CROPS (GREENHOUSE)

- ★ Lettuce
- ★ Peppers
- ★ Tomatoes

* ★ Cucumbers
* ★ Herbs
* ★ Flowers

CROPS (FIELD-ANNUAL)

* ★ Asparagus
* ★ Beans
* ★ Carrots
* ★ Broccoli, cauliflower, Brussels sprouts
* ★ Lettuce
* ★ Peas
* ★ Corn
* ★ Pumpkins

GRAINS

* ★ Corn
* ★ Hay
* ★ Alfalfa
* ★ Straw
* ★ Canola
* ★ Oats
* ★ Wheat
* ★ Barley

NURSERY CROPS

- ★ Evergreen
- ★ Deciduous shrubs
- ★ Flowering trees
- ★ Fruit trees

MUSHROOMS

- ★ Exotics
- ★ Button

CHRISTMAS TREES

FRUIT TREES

- ★ Apple
- ★ Peach
- ★ Pear
- ★ Cherry

FRUIT CROPS

- ★ Strawberries
- ★ Raspberries
- ★ Blueberries
- ★ Cranberries
- ★ Blackberries
- ★ Logan berries
- ★ Grapes

ORGANIC FRUITS AND VEGETABLES

ALTERNATIVE CROPS
★ Garlic
★ Medicinal herbs
★ Culinary Herbs

BEES
★ Honey
★ Beeswax related products
★ Beekeeping and tending

MAPLE PRODUCTS
★ Syrup
★ Candies

AGRICULTURAL TOURISM
★ Bed-and-breakfast working farm
★ Farm tours
★ Corn mazes
★ Pumpkin patches

FARM SERVICES
★ Contract spraying
★ Contract baling
★ Custom tractor work

After reviewing the above possible opportunities, consider which might best suit your interests, skills, abilities, and local market opportunities, and begin devising a list of all possible options. To help in narrowing that list, there are a few important factors that might make certain commodities more feasible to include in your farming operation than others.

EASE OF ENTRY INTO THE MARKET

As a new farmer, or a farmer looking to expand an existing operation, you need to consider how easy it will be to enter the market for a particular commodity. The easier it is, the faster it will be for you to get started – but at the same time, the faster it will be for competitors to enter the market and potentially drive down the price of the product.

Consider Christmas trees. Many people choose to grow Christmas trees because it is relatively easy to get started. The problem is that the prices the trees are able to command are much lower than you might expect: The market has become flooded due to the ease of entry into the Christmas tree market. This is why many Christmas tree farms have branched out and opted to offer other services, such as hot chocolate and Christmas cookies, sleigh rides, Christmas ornaments, etc.

EASE OF EXIT

You must also consider the ease of exit for that particular market. If for some reason your competitive position should begin to erode in the market and you opt to change to a different commodity, you must determine whether the capital that has been invested in producing the original good is so high that it would simply be too expensive for you to make a change. You do not want to find yourself in a position where you are locked into operating at a lower profit than you originally planned.

MARKET SIZE

You must also always consider the size of the market for that particular product. Make sure you are objective in this because it is extremely important to your overall success. Many factors must be considered when estimating the size of the market for any product. Ask yourself these questions about any product you are considering producing:

- ★ What is the size of the target market for the area I want to serve?
- ★ At what price can I sell my product?
- ★ What percentage of the market can I realistically capture?

In terms of producing a commodity that is fairly traditional, you might begin by identifying the per capita consumption of that product for your area and then multiplying that by the population of the area. Keep in mind that one of the most common mistakes that many people new to the idea of small farming make is in overestimating the size of the market they can realistically capture. You should always consider both best and worst-case scenarios.

It can be difficult to determine the price you can realistically charge for your product. If you lower the price, you very well might be able to sell more but you will have a lower margin. Conversely, a competitor might lower the price to meet your price, and then you will be selling at the same price and still have lower margins. There are three approaches that can be used when it comes to pricing:

1. If you are competing with another farmer on price, you might try to have the lowest price for a particular quality.

2. If you are selling a product that is unique, you might try to charge whatever the market will bear. The key here is the uniqueness of the item and how it's marketed. If consumers perceive that it is of greater value, they may be willing to pay a higher price for it.

3. If you are producing and selling a product that is similar to others in the market, you may need to price some of your items at slightly lower than the market value so that customers will make the switch from the competitors to you. It is up to you, then, to keep your customers happy so that they buy other items from you, hopefully at full market value.

WHAT AM I ABLE TO PRODUCE?

At this point, you may have some ideas about what you want to produce and would be able to command a place within the existing market for your area. Now it is time to take a closer look at whether you will be able to produce those products competitively on your property. This involves assessing the capability of your land and matching the characteristics of your land with the requirements for production.

CHAPTER 2
LIVING OFF THE LAND

"When tillage begins, other arts follow.
The farmers, therefore, are the founders
of human civilization."

– Daniel Webster

LIVING OFF THE LAND

At 93, my Great Uncle Bob McNinch speaks from the experience that only decades of farm life can ingrain in the soul. Bob is three years younger than his brother, my grandfather Jim, and they both grew up and farmed near the small town of Belmont, N.Y. Bob is still living there with my Great Aunt Beth today. They are the only two farmers of the seven brothers and sisters and have outlived by at least ten years all their other siblings. Kind of makes you wonder what role farming played in their longevity.

Uncle Bob has one of sharpest minds of anyone I know. He is a history buff and can fashion an interesting story on just about any historical fact you might care to discuss. One of my earliest memories was going to his farm while he was boiling sap to make maple syrup. That sweet aroma stays with you forever.

In fact, before I say anything more about land use and soil conditions and whether to buy or lease a farm; let's hear his story, rich with anecdotes, in his own words:

UNCLE BOB'S FARM DAYS

In 1923, my family moved from the farm where I was born in 1919 into a new house in the town of Belmont, N.Y. A field behind the house extended to the river, and this field was big enough for a large garden and a pasture of two or three acres for our dairy cow, Diantha. The produce from the garden and the milk from the cow provided a lot for our family of seven children.

At times the river could overflow and take out the pasture fence. One time my brother Jim and I were given the opportunity to dig the flood trash from the downed fence, dig out the wire and fence posts, and put the fence back in place. Another effect of flooding would be a new crop of weeds. One year the pasture field grew up in wild carrot. My dad let Jim and I pull the weeds. We made several large piles, and burned them. There was a side benefit of this job—we were not far from the local swimming hole. (At that time, one didn't need a bathing suit to swim in the river.) Sometimes we went to the other side of the river and had a snack from Dr. Howe's raspberry bushes.

Some summers my dad would pasture Diantha and her daughter, Susan, behind my grandfather's barn across town. Jim and I would walk over every evening to milk the cows. I'd milk Diantha. Jim, being the older brother, would milk Susan, since she was a little more fractious. Jim would take the pail of milk from Susan up the hill to our grandfather's house, and then we would walk home together. Usually we kidded around some, jumping out from behind trees to try to scare one another, or locking the pasture gate while the other was still inside.

One night as Jim took the milk up to the house, I suspected he had locked me inside as usual. So I took the pail of milk from Diantha and climbed over the back fence, figuring I could fool him that way. Jim and I met at the corner, and we went on home. In the morning, my dad got a call that the cows had gotten out and a train had hit Susan. The train crew buried her by the side of the tracks. I never went home over the back fence again.

Over Easter vacation during my junior year in high school, I rode my bicycle down to the Belvidere Farms to apply for work for the summer months. I was told to come back when school was out. The Monday after school ended, I went to work. I was 15 years old, and the pay was one dollar a day. My dad took me down to the farm in the morning and picked me up after 5 o'clock.

My first job was to help make hay. Another fellow and I went to a field, hooked up a wagon and followed a raked windrow of hay. My job was to load the hay on the wagon.

The other fellow drove the horses and stayed out of the way. I finally asked him where the rest of the crew was. He told me it was just us two.

Another one of my first jobs was to take the horses and wagon and a pile of feedbags to a sawmill on the back road. I had to fill the feed sacks with sawdust, load them on the wagon and bring them back to the farm. It was quite a responsibility for a kid who hadn't driven horses much.

I also took care of a large flock of chickens. One rainy, cold afternoon, when my work was nearly done, I put some fresh eggs in a tin can with some water, put a blowtorch next to the can, and boiled up some eggs. They tasted pretty good but needed salt.

The farm manager was Frederick "Fritz" Volmar. He had been an officer in the German army during World War I. He was a strict taskmaster, but fair, and I learned good farm practices from him. He often would give me a job and let me figure out the best way to accomplish it. I learned to repair and adjust the mantles on the gasoline lanterns that lit the henhouse in the evening. I was educated in the art of shingling and how to use a board to keep the shingles straight as I put them on. I learned how to harness and drive horses and the most efficient way to clean out a barn.

Mr. Volmar gave me a lesson in planning, too. In the fall we were getting ground ready for sowing wheat. I was given the job of dragging a plowed field, with a pair of horses to pull the drag. I saw Mr. Volmar drive to the edge of the field in the farm car, stop, and wave his arms, motioning for me to come.

It seemed he wanted me quickly, so I unhooked the drag and hurried with the horses to the gate where he stood. When I got there, Mr. Volmar told me it was lunch time – and then made it quite clear that unhooking the drag had made no sense. The horses could have pulled it across the field. "You could have had a free trip," he explained. I was embarrassed, but I have thought of this many times over the years – and taken the advice he gave me many times, too. I worked on the Belvidere Farm during summers and after school until December of 1936.

When my grandfather James Hall died, my mother inherited two farms. There were tenants on these farms, and the custom was to work a farm on "shares." The owner would furnish the farm, cattle, buildings, half the feed, and get a share of the milk check. The tenant would furnish his equipment and plant and harvest the crops. In December 1936, the man working one of my mother's farms told her he would be leaving at the end of the month. My mom asked me if I would be able to fill in until they could find someone else.

It was a challenge, but I thought I could do it. I'd had a year or so of experience working on a good, well managed farm – and, what the heck, I was almost 18 years old!

One of the first things I needed was a pair of horses. The tenant's father said he had an older pair of horses that he would sell, with harness, for $150. When I saw them coming up over the hill, I hoped they would be going right on by. They didn't look too good, but Molly and Dan turned out to be the right ones for me to start with.

After a year or so, Dan developed kidney problems, so I bought Prince, a western horse. He wasn't used to farming, so it took us a while to get used to each other.

Among the lessons I learned was the importance of record keeping. One threshing season, I asked for help from a fellow who had worked for me the previous year. He informed me that I hadn't paid him the year before. I sorted through my canceled checks – and there it was, the check I'd given him, cashed.

Another lesson came along a bit later. From another farmer, I bought a registered cow to add to my herd. I hoped to be able to someday have my entire herd made up of registered cattle. It wasn't long before I found out that this new cow had brucellosis. It spread to all the pregnant cows in my herd; they all lost their calves, and I almost lost my livelihood. It took a long time to rebuild the herd and I learned a valuable lesson in vaccinating all new calves. I eventually built up a herd that was resistant to brucellosis. My mother never did look for someone else to take over the farm.

In the spring of 1939, a young lady named Beth Huey came to Belmont to apply for the Home Economics teacher position at the school. My sister Louise was a freshman in high school and was in Miss Huey's class.

My mother, who was on the Board of Education, was heard to remark, "We like to have young ladies come to Belmont to meet our young men." Sometime in the early fall the new teacher received an invitation for an evening meal at the McNinch home.

My mom woke me up from my nap on the couch after the meal and asked me to take our guest home. This casual meeting led to a second meeting. On a Sunday afternoon, I borrowed my dad's car to take the young lady for a ride. My own vehicle, a farm pickup truck, was not suitable for a Sunday drive. However, I did not check the gas in Dad's car, and somewhere near Belvidere the car quit. I had no money, but my companion did, so I borrowed fifty cents to buy enough gas from a neighbor to get us home. Beth saved the day! After a few more meetings, we came to an agreement to get married in July 1941.

We needed a suitable house. With some misgivings, I showed her the farmhouse on the land that I had been working near Belmont. Nobody had lived in it for some time and there was a big hole in the dining room floor where the furnace grate had been taken out. There was no electricity, no cupboards in the kitchen, and no bathroom. A large spring provided water, but it wasn't piped into the house.

"Well," Beth said, "it doesn't look too bad."

We began getting the house livable. Nearly everyone in the family pitched in. My mom spent evenings helping wallpaper the rooms. We shone the headlights of the car into the house for light. My brother Louis painted some of the floors. Beth's mother cut up old wool Civil War uniforms and made an 8 foot braided rug for the dining room. (We later found out that these uniforms were quite valuable; we probably could have sold them and carpeted the whole house.)

She also refinished an oak extension table and caned and refinished several chairs. Beth restored a davenport by reupholstering and retying springs, and her dad built a 4-shelf bookcase for us.

The electric company assured us we could have service by the middle of summer. Before that, we had to have the house wired. To get the wiring from one side of the house to the other, my brother and I removed a floorboard on each side of the house. We snagged a barn cat, tied one end of a long piece of string to her tail, and pushed her down through one of the floorboard holes. On the opposite side, we shone a light through the hole and called the cat, who rushed right over, pulling the string with her. We untied it from her tail and then tied the electrical wire to the string, which we used to pull it through.

Arling Saunders already had removed the wood-burning furnace from the house but hadn't installed it yet at his place. He was willing to let me buy it back. I traded a stack of hay that had been put up the year before. Beth used her home economics training and drew up plans for kitchen cupboards. I was not a great carpenter, but the completed cabinets, new sink, and bathroom looked pretty good. The woodwork was repainted, and I built a corner cupboard in the dining room for dish storage.

With family help and lots of hard work, the house was ready for us to move into after our wedding. Two weeks later, they turned the electric on.

In December 1941, most of the young men from Belmont went to Buffalo to enlist in the armed forces. I made it through the physical, but when they asked me what my job was, and I said "farmer," I was sent right back home. Farmers were needed on the home front to raise food for the war effort.

Beth and I began improvements on our farm. I purchased a registered bull calf and a few heifers from a local registered Holstein breeder. This was the start of eventually having a registered herd of 50 milking cows.

We made good use of the Extension Service and the Soil Conservation Service. We added a poultry business that enabled us to have a daily income from the sale of eggs to supplement the monthly milk check. We planted several thousand evergreen trees on land that was difficult to cultivate, and we had five ponds dug for migratory wild fowl and fire protection. As we added acreage to the farm, we had diversion ditches cut to allow strip cropping on hillsides. At the completion of that last project, our farm was chosen as the Conservation Farm of the Year.

Over the years, we were able to add to the farm buildings. These days, you'd call it recycling: I bought a large unused barn a few miles away for $50. It had a lot of good timbers in it. I removed all the inside bracing, hooked a rope to the end peak of the barn and used my Brother Jim's tractor to pull it down. The boards flew right up in the air, and about all we had to do was pick them up.

With the materials and laminated rafters, I erected a 25'x75' two-story building, with the second floor used for housing a laying flock of 600 chickens and the first floor used as a granary, work shop and storage.

I also bought a barn in town, tore it apart, and rebuilt it as a storage barn and upstairs granary. Another barn I tore down gave me material to add 16 feet to the existing dairy barn. Later we extended this barn further by using wood we had cut on the farm. We moved in 3 brooder houses for baby chickens, too.

By steps we also increased the farm acreage. We bought 50 acres of land adjacent to our existing farm, and 70 acres of crop land on another side of the farm. We also bought part of a farm that included some good timberland on the road back of our farm, and we added all of a 160-acre farm with a good house and barn.

By the mid 1960s, we were farming 560 acres and had a full-time employee who lived with his family in the other house. By working hard together, making plans, asking for advice, reusing materials as we could, and taking advantage of various government programs, we gradually made improvements. We were able to live a good life on the farm.

— Robert McNinch, Belmont, N.Y.

YOUR LAND CAPABILITY

In determining which types of crops or livestock would be best suited to your property, it is important to analyze your land, including the type of soil. This will help you to eliminate certain commodities as well as better understand the types of management practices that may be needed to overcome soil problems.

Developing healthy soil and pastures is part art and part science and is a critical component to most farming enterprises. If you're starting with poor soil, you will need to factor in the additional expense of adding the necessary inputs to improve its condition. Ultimately, the rewards of being a good steward to your soil will show up in the crops and livestock that you raise.

TOPOGRAPHICAL RESTRICTIONS

The topography of your land can restrict you from raising certain crops or can require more complex management practices.

Is your property flat or hilly? Hilly land can be ideal for raising livestock, but if you want to grow crops, the topography is going to add to the costs of production.

If your property is sloped, do the slopes face north or south? This is important because light intensity tends to be greater on southern slopes, which can be important for certain crops.

Is any part of your land wooded? If you are interested in growing any type of crop that requires shade, such as certain types of ornamentals or shiitake mushrooms, you must have shade. Wooded land can provide the shade you will need naturally. Otherwise, you will need to build some type of artificial shade structure.

Do you have any large, open, flat fields? If your goal is to grow large volumes of field crops, you will need to have fields that are suitable for this type of production.

DRAINAGE AND SUITABLE WATER SOURCE

Are there any poorly drained areas on your property? Land that is poorly drained can reduce the productivity of certain crops and ensure that you get your tractor stuck more than once. You may need to add some drain tile to solve certain water issues. I think we have run over 2,000 feet of drain tile on our property so far to handle poor drainage areas and to control other watershed issues.

Do you have a good source of water for irrigation or animal watering purposes? We learned that carrying five-gallon pails of water between barns in the middle of winter was no fun.

The following spring, we put in a new water line and an outdoor hydrant near the barn where we keep some of our alpacas. What a relief!

HEALTH OF SOIL

What is your soil like? This can prove to be vitally important to the success of your crops. In most instances, you will need soil that is rich and well-drained to successfully grow a crop. To help determine what type you have, you can usually get soil maps from your local soil and water conservation office. Our soil has a lot of clay in it and needed a lot of compost and attention during the first few years to allow for some good yield from our garden. Our alpaca manure is an ideal fertilizer for which we have a never ending supply. We have several gardeners who swear by it and come back year after year to get a truckload.

GROWING SEASON

What is the growing season like in the area where you live? How many crops will you be able to realistically grow in one season? Resources to help you in this pursuit are:

U.S. Department of Agriculture: **www.usda.gov**

Sustainable Agriculture Research and Education:

www.sare.org

OTHER IMPORTANT FACTORS

You should also consider whether you can protect your crop or livestock from theft and whether your land is easily accessible to the public. The latter might be important if you are interested in having customers come to you. If your land is remote, you must have plenty of signs to direct customers to your location. Even with today's GPS navigation systems and online maps, people are people and will surely get lost. In addition, consider aesthetic values associated with your property. If you are planning to offer a rustic experience or a farm tour, you should have land with rural beauty and an inviting homestead.

EQUIPMENT AND OTHER ASSETS

What equipment will you need? If you do not own any such assets, you will need to factor them into your overall small-farm business plan and make arrangements to purchase what you will need.

For example, do you have a tractor? What about a greenhouse? Both are essential for growing certain types of crops. Early on, one of the best decisions that we made was to purchase a tractor. My Kubota B7800 is a workhorse, and we could not farm without it.

From cutting grass, to drilling post holes, to using the bucket and back blade to move earth, poop, sand, hay and snow, it is an indispensable asset.

What about a barn? If you do not have a barn, you will need some type of shelter where you can store or house livestock, tools, supplies, seeds, and your harvest. We have our main 30'X60' pole barn with twelve-foot overhangs that house our hay, animals, and tools. We also have several out-buildings and a greenhouse.

Do you have facilities for washing, cooling and drying? A lack of such facilities can greatly affect the success of many crops. For instance, if you plan to grow herbs, you will need a shelter for your dryer that will be suitable for your particular climate and crop.

ACQUIRING FARMLAND

If you are considering starting a small farm, you may already have your own land on which to farm. Or you may be in a situation like we were with no land and are looking for that ideal property to start your farming venture. If you do not have your land yet, you will want to carefully consider whether to buy or lease your land. We decided to buy since we were going to move to the property once the farmhouse was remodeled. Farm land in our area is not cheap and we hoped that by purchasing the land we could build some equity for the future.

For many though buying land may not be the most financially prudent way to go. You may want to lease or rent to get started.

LEASING FARMLAND

Land in many areas of the country is getting more expensive to own. Some landowners and older farmers would appreciate having a piece of property that is now sitting idle returned to some type of farming enterprise. The landowner can earn some money, helping to offset taxes, and you can gain access to land that you may not initially be able to afford.

Another advantage of leasing is that you can end the lease if your farming desires begin to wane or you change your mind on the direction of your farm. You can move on and not worry about having your capital tied up in land that you may not be able to sell quickly.

Leasing is also a good option for farmers who are looking to expand their enterprises. The actual arrangements for leasing can vary a great deal, based on the length of the lease, the agricultural value of the property, the market value of the property, and the prospective costs and benefits to the lessee. To avoid misunderstandings, you may want to have an attorney or other qualified source draft a proper agreement stating what is included in the lease agreement. This protects both your interests and may save a legal battle down the road.

As you have seen from the preceding discussions, there are many factors to consider when looking at procuring the ideal farming property. Understanding the necessary requirements can help you to select a property that will be best suited for your chosen product or commodity.

CHAPTER 3

THE WAY WE GREW

"Farming looks mighty easy
when your plow is a pencil, and you're a
thousand miles from the corn field."

– Dwight D. Eisenhower

THE WAY WE GREW

S mall farming today is worlds apart from the homesteading days of our forebears. Those were the days of subsistence farming. Today, it's essential to choose a farming pursuit that will return enough profit for your subsistence.

You need a high-value crop, one with potential to make the most from small acreage. In fact, for more and more farmers, the "crop" isn't a crop at all – and the growth of agritourism, in which city folk and their children come to experience the farm life, is but one reflection of how rural life has changed.

We'll talk more about value crops later in this chapter, but first let's hear about values – the ones we grew up with, the ones we want to preserve forever by getting back to the soil. Let's hear some reminiscences from those who have lived the changes in small-scale agriculture, as time and technology altered forever how we did things.

THE ARTIFICIAL INSEMINATOR

It was my cousin Dave who found the thing. It was lying in the fading light at the edge of the woods, hidden amid the skunk cabbage in a swamp pocked by cow hooves.

Sloshing through the muck in gum boots, we gathered to inspect his finding. We stood silently around it in a circle, five boys in yellow rain slickers – my three cousins, my brother and I, in a tableau frozen in time, Mercer County, Pennsylvania, June of 1966.

"Gawd, look at that rope coming out of its head," Dave said. "And check out that eyeball!" We dwelled on every feature as the first heavy drops of a storm pelted away the horse flies from the glistening body of the calf fetus.

I allowed that it was the coolest thing I'd ever seen, but I wouldn't sleep well that night, unable to shake the image of a cyclops face with an appendage that looked like a tail growing from its forehead.

My father, who had been a dairy farmer for decades, had sent us out to the woods that evening after one of the Holsteins from his small herd had come plodding into the barnyard at milking time with a ribbon of afterbirth trailing behind her in the mud.

Our job was to find her calf, as we had done numerous times before when a pregnant cow failed to return at milking time.

Sometimes we'd search for hours, peering beyond the rim of light cast by our hissing Coleman lanterns. Usually, though, we'd be looking for the mother, too, who instinctively would try to secret her newborn into the deepest of the blackberry thickets.

Dad knew something was amiss this time. The mother had left her baby alone in the woods. And we had found the stillborn freak right where she had expelled it, unceremoniously, perhaps pausing to sniff at it before heading homeward toward the comforts of the stables, dispirited yet ready for a scoop of grain.

It was time to get rid of the bull, Dad proclaimed the next morning up at the farmhouse over pancakes and sausage. Too much inbreeding caused monstrosities like that. We would go modern: The next time a cow went into heat, he would call the artificial insemination service.

I didn't care much for the bull anyway. For years it had ruled a building attached to the barn – "the manure shed," as we called it, a holding pen beneath the straw mow overhead. Mostly the old bull stood there chewing its cud, looking dazed, until my father led a willing cow into its domain and tethered it to the heavy wooden gate. "They're playing piggyback!" I thought as I watched, with the eyes of a 5-year-old, perched nearby astride the fence.

It was the most energy I ever saw the bull exert, though my cousin Craig was sure that he once had narrowly escaped a good goring. One day after supper, we children were playing in the mow, diving from a beam into the mounds of straw, when Craig vanished through a hole in the barn floor. Finding himself down in the manure shed, he never stopped to see whether the bull even noticed him but dashed through a crack in the gate, ripping his belly. He opened his belt to show us the gash, and then cinched it under a wad of burlap. "Don't tell anybody!" he beseeched us, as if someone might blame him, and never, till now, did I dare.

His flight of terror had left him with an injury far worse than the bull ever could have inflicted. Its piggyback days were drawing to an end.

Watching an artificial insemination was even cooler, for a boy, than finding a freak fetus in the rain. Dave and I leaned against the whitewashed walls behind the stanchions and watched the technician at work.

Instead of leading the cow to a romantic rendezvous, my father had selected a semen donor from a catalog of fine muscular specimens. A young Holstein heifer would get an Angus mate, for example, which produced a small calf – easier for a mother giving birth for the first time, yet sure to grow fast for the auction. I did not know, nor do I know now, nor do I want to know how the semen was obtained from the bull, but the technician produced a vial of it, and we watched as the young man set to work.

A cow in heat is a needy creature. She will try to mount nearly anything, not understanding the correct position but just knowing she needs to do something, anything, to satisfy that urge. She will mount other cows in the barnyard.

Once, as a teenager, when I knelt to move a water bucket from in front of a cow in heat, she reared up in her stanchion and tried to mount me. I didn't tell anyone about that.

As any good artificial inseminator knows, it just won't do to inject the semen and run. A cow needs some attention.

As we watched, the technician pulled a long latex glove onto his right arm like an immense condom, pushed aside the tail, and thrust his arm deep into the cow, thrusting rhythmically. The cow's back arched; her eyes bulged. The technician inserted a tube, pushed a plunger, and went out to the barnyard for a cigarette.

Dave and I looked at each other, and later we would compare notes on just what we had seen and whether it had been so. I do not recall how I recounted the episode to my buddies in school the next day. I'm sure I did.

I suspect, however, that the young man with the latex glove didn't have much to say about his day when he went to the tavern that evening. I doubt he regaled the ladies with tales of his prowess on the job. And somehow it seemed that whenever a cow came into heat, a different young man arrived to do the deed. "It's what you would call," my father explained, "a job with a high turnover."

– Bob Sheasley, Collegeville, Pa.

Now let's meet another farmer, who learned a related lesson both in keeping with the times and the importance of being neighborly. Farmers must deal with all manner of troubles and not get bent out of shape when things get tough – or rather, as you will see, they must stay tough when things get bent out of shape.

A LESSON LEARNED

The importance of helping neighbors is unparalleled – especially in an agrarian community.

Over twenty years ago, my wife and I started a breeding business for registered Hereford beef cattle. The business was actually a continuation of the family farm on which I had grown up. We finally had the opportunity to purchase land, a house, and a run-down barn.

We had assembled a group of cows, and now it was time for a bull. A longtime friend told me about a wonderful genetic package that was for sale in Maryland. We purchased the bull, named Victor, and brought him back to our farm in the Penn Yan/Branchport area of New York.

Cattle people marveled at this bull. He was a true beauty. His first calf crop hit the ground about nine months after we turned him in with the cows. The calves were everything we had hoped for.

My neighbor, who had helped me tremendously during our formative years, asked if he might breed his virgin heifer with Victor. I was hesitant – but how could I say no? After all, Steve is a wonderful man who still lets me bale hay on his land and hunt deer in his woods.

I went up the hill and loaded his heifer onto my trailer. She was a true prize – a young thing of four years that his daughter had purchased for him. The heifer was almost as tall as my belt buckle, and about as wide as she was tall. Steve was proud of that heifer. He had been considering breeding her, and since I had this new bull in town, he figured he would ask.

I backed my trailer up to my pasture, and off-loaded the animal. The rest of the work was up to Victor. I did not see a lot happening between the two, until it was too late.

One day after I came home from teaching school, I found Victor standing in the field with a condition. The condition was like no other in agriculture. It stemmed from the physics of a bull mounting a cow.

Victor weighed about 1,800 pounds. The heifer weighed about 700 pounds. When he mounted her, she collapsed under his weight. However, Victor's penis was still inside the heifer when she went down, and it bent at about 90 degrees.

That's when I learned what a "broken" penis looked like. The immense swelling did not allow Victor to retract his penis completely. This caused him another series of issues.

We loaded the bull onto my trailer and headed to Cornell, where Victor had all of the attention and medical assistance one could hope for. I don't remember how much the vet bill was upon his discharge, but I do remember that the veterinarian told me that my job would be to keep his penis clean.

"You'll need to keep it washed," the vet said, "and here's an ointment to repair the skin. Apply it daily."

This was the longest winter in my life.

Victor's prognosis looked good. Everything shrunk to normal, and it appeared as though Victor would have a continued life of breeding.

Alas, in the spring, we turned him out with the cows and waited. With great sadness, we saw that the sparkle in Victor's eye was gone. It seemed he could care less about breeding cows. He found it far preferable to lie under the trees rather than take that risk again. His heart wasn't in it. Victor's life as a service bull was over.

The lesson here is simple: Continue to help your neighbors, and continue to select superior genetics to use as breeding animals. However, the simplest way to preserve genetics is to have the bull's semen collected and frozen. That way, if anything ever happens to the bull, you will forever have his genetics.

The real kicker of this story is that we had an actual date to have that procedure done – right after he finished breeding my neighbor's heifer!

– John Kriese, Branchport, N.Y.

SELECTING HIGH-VALUE CROPS

By their very nature, most small farms are limited in resources. What you lack can determine what you will be able to do with your farm.

The land itself is one of the biggest factors limiting your pursuit. Livestock and crop productions that have high overhead expenses and a low unit return are not usually going to be profitable on a small-scale farm where acreage is limited. Some livestock options, including beef cattle operations, can be produced on a small scale profitably, but you may find that the income alone is not enough to sustain the farm.

The cost of purchasing land will have a direct impact on your ability to make a profit from the farming operation. We live near an area where good crop land is selling for over $5,000 per acre. In order for the farm to be profitable, you must generate more than the total cost for both the variable and the fixed costs associated with the operation.

The best chance for making additional money for many small farming enterprises is choosing a crop that is high-value, particularly with niche and specialty markets. If you can position yourself to be among the first producers to market for a particular crop, you have the opportunity to set the price for it. Once competitors enter the picture, the market will begin to control the price for the crop.

Below we will review some potential high-value crops you might consider. You should always consider which crops can be realistically and profitably grown based on your region, your land, resources, skills and abilities.

SPECIALIZED NURSERIES

One option to consider is running a specialized nursery selling plants either retail or wholesale. You might consider selling wholesale to garden centers. Another option would be to operate an internet or mail-order nursery. You could also sell directly from your farm.

AGRITOURISM

Have you ever considered having tourists and children come out to visit your farm? If you like working with people, this very well could be an opportunity for you to turn your farm into a tourist destination. You could entertain and educate while selling your products at the same time. Many farms near us have embraced this concept and have added pumpkin patches, corn mazes, petting zoos, and farm events such as a craft fair with multiple vendors or other entertainment to help bring in additional revenue. We have had over 100 people show up on a Saturday during an "Alpaca Farm Days" event during our first few years in business.

This helped to get the word out about our farm and let's people get introduced to the alpaca on a personal level.

ORGANIC ANIMAL LIVESTOCK

The continued debate regarding the safety of the meat supply in the United States has led to the development of a strong market for meat animals that are grown organically. Premium prices can be commanded for organically grown sheep, beef, rabbits, poultry, and other animals. The USDA has been developing standards for certifying organic meat. The demand and market for free-range products, including eggs and poultry meat, has been very strong. There has also been an increased demand for meat goats in many areas around the United States, particularly those areas with larger ethnic populations.

ORGANIC HAY AND FEED GRAINS

The organic food market has exploded in popularity in recent years, and as a result there has been an increased interest in producing organically raised livestock. This in turn has led to an increased interest in organic feed for livestock including horses raised by owners with concerns about the safety and quality of the current feed supplies.

The lack of a consistent and reliable source of organic feed has caused many producers to avoid the organic livestock market. The development of organic feed production would involve an initial capital investment due to the fact that the farm would need to be certified through an organic certification program. Even so, the premium that can be achieved for organic feeds and hay could offset such an investment over time.

ORGANIC VEGETABLES AND FRUITS

Fruits and vegetables that are grown organically are always in high demand and can command an organic price premium. The key to succeeding with this opportunity is producing a product that is not widely grown in your local area or developing a marketing strategy that would persuade customers to purchase your product over a competitor's.

UNUSUAL OR HEIRLOOM VEGETABLES

There is also a strong market today for vegetable varieties that were popular in the past. Along with heirloom varieties of vegetables, there is also potential to develop a niche market for vegetables that are uncommon.

Such unusual vegetables include white eggplants, blue potatoes, etc. As is the case with other options, the key to succeeding with this is to start before anyone else in your area begins growing these crops. I have listed a few seed companies in the index to help begin your search.

GRAPES FOR WINEMAKING

Wine grapes already provide a strong market potential in many areas due to the presence of wineries. In the Finger Lakes region of New York, this has become big business. Many varieties of grapes can be grown to provide a niche for a start-up farm looking to sell grapes to local wineries or via festivals and roadside stands.

ORIENTAL VEGETABLES

The potential to obtain a high value for oriental vegetables is excellent, particularly in areas with ethnic populations. A wide variety of oriental vegetables and fruits can be grown. You would need to research which varieties grow best in your region.

HERBS

The production of medicinal or culinary herbs also presents an excellent niche market in many areas. Along with the potential for marketing dried and fresh herbs in your area, you also could market your herbs on a wholesale level.

DRIED AND CUT FLOWERS

The market for flowers that are locally produced to be cut and dried is also quite strong. If you happen to have country stores or direct markets in your area, you may find that this is an excellent opportunity. Remember to select varieties that can be easily grown in your climate and region.

ORNAMENTAL SHRUBS AND TREES

Urbanization in many areas has created a strong market for ornamental shrubs and trees. Keep in mind that production of such crops can require a large investment of time and capital; however, the profit potential is also large.

GREENHOUSE OR HOOP HOUSE

The production of off-season vegetables and bedding plants is growing in popularity. These wonderful structures allow for an extended growing season in many areas of the country where colder weather limits the time that crops can ordinarily be grown. Many farmers' markets and CSA (Community Supported Agriculture) programs have been able to extend their season, allowing consumers a longer period in which to purchase locally grown produce. Depending on resources you already have, some higher start-up costs can be associated with this type of production, including the construction of a greenhouse and the need to heat it. A simple hoop house may help to lower your costs. The financial rewards that can be achieved through raising greenhouse crops can help to offset those costs.

HYDROPONICALLY GROWN CROPS

You might want to explore growing some of your produce such as herbs, strawberries, tomatoes or lettuce hydroponically. This essentially is the growth of plants in a nutrient solution without soil. You will have higher potential startup costs with this type of operation but you may find the financial rewards worth the effort.

The need to select a high-value crop so that you can make it as a small-scale farmer has developed through generations of change in agriculture – and in our lifestyles. The small farm that many still remember is slipping into history in America, due to technological and economic pressures that have changed our culture in many ways. Lest we forget, let's look back. Here are two accounts of farm life in the past century. These reminiscences say more than any sociology textbook or dissertation on the evolution of agriculture could ever hope to explain:

FARM LIFE IN THE 1930S

The Pulver family farmed in Italy Hill, New York, for three generations on Pulver Road from the 1840s to the 1940s, beginning with George, who was born in Kinney Corners. By the 1920s and 1930s, my Pulver cousins and I were the third generation born in the same family home. We spent summers and Christmases there – and Thanksgivings, which usually saw fifty or more close relatives sharing food and togetherness.

The long summer days of our childhood were filled with whatever we could find to amuse ourselves.

No radio, TV, or other electronic devices. There was a lot of "make believe" and pretend. I remember riding tricycles with my "almost twin" cousin on the large front porch. We'd set up potato crates and bake mud pies, and from the garden came an assortment of vegetables. Grandma Mary would can over 400 quarts of fruit, meat and vegetables each year, and more of the harvest was kept in cold storage in the cellar.

The maple trees in the front yard were a delight for agile children. We would climb higher and higher. Our swings were two tires hanging in the back yard. If a nickel was somehow available, we would walk the two-mile round trip down to Brown's Store to purchase a package of Kool-Aid to make a treat with the cold well water.

The only adventures outside the farm were on Sunday, when we would walk down to the Baptist Church for Sunday school. On returning home and passing the Kennedy home, the three maidenly sisters might ask us in for some lemonade, which was a very special treat for three young children. We also would go to Prattsburg with Uncle Walton in his big old car to get supplies, such as chicken feed. The big sacks would fill the back seat, and we would ride on top of them, with no thought of buckles, on the way home.

We always had chores. Eggs had to be gathered, which I liked to do, but not cleaning out the coop. The cows who spent the afternoon in the shade right next to the barn would invariably wander off about two hours before milking time, and we would have to round them up in the field farthest from the barn.

Grandpa had a tractor with lugs, no tires, and the haying was always done with the horses, Bess and Bell, hitched to the wagon while the hay was loaded by pitchfork. It was hot, dirty work for sure, made even more unbearable by the ever-present "sweat bees." Our chore was to carry jugs of cold water up the road to Grandpa and Uncle each morning and afternoon. No such thing as ice or insulated jugs. Later, the horses pulled the hay up to the hayloft, and our reward was free rein to jump into the new hay. No one worried that we would fall or get hurt – it was just what we did.

Modern-day conveniences weren't present either, but Grandpa did have a generator so there were lights most of the time, which was a great help. It was around 1940 before electricity came to Pulver Road. It was always the "outhouse" for all those years, but we had the best. A three-holer! And a good thick Sears catalog for toilet paper. Grandma had her elderberry bush outside the outhouse, and it grew very well, perhaps not surprisingly.

Life wasn't easy in those "good old days" – but certainly it was a rewarding family life with joys, sorrows and many hardships. I feel a certain sadness that today's generations cannot experience those times. I am grateful that I have known those other days and other ways.

– Nancy Diven Gillette, Penn Yan, N.Y.

After the Great Depression and World War II, agricultural and societal pressures continued to shape the small-farm operation. In the following account, you will see a glimpse of life as it came to pass in the ensuing decades:

FARM LIFE, LATE 1950S AND EARLY '60S

From the magnificent perch of my grandparents' farm in Steuben County, New York, one could gaze down through the purple hue of the hills into Pennsylvania. The farm was in Highup, appropriately named: At 2,400 feet above sea level, it is the highest point in the county, midway between Greenwood and Troupsburg.

I was quite accustomed to town living, growing up on a side street in the small village of Nunda in Livingston County, where I could walk uptown, get ice cream from the soda bar, go to the playground, and ride my bike on the sidewalks. I enjoyed village living, but my lifelong favorite memories always take me back to that farm near the state line, where my cousins and I spent our summers and Easter breaks.

The 800-acre dairy farm was the home of my mother's parents and is still in the family today. The land is about half wooded and half tillable. My grandparents, Joe and Maude White, lived in the main house. My aunt and uncle, Anna and Wilfred White, also lived on the farm, about a quarter mile down the road.

All four of those fine folks worked on the farm. My grandmother was best at canning, pickling and providing meals for us field hands. She would either have lunch and dinner ready at the house or bring it to the field where we worked daily. My aunt could do most any field task such as drive tractor, load hay on the wagons from the baler, and carry milk, along with her other household duties.

My grandfather gave me my first lessons on how to drive a tractor while sitting on his lap and shifting gears. He taught me that when you ran over a nest of ground bees while haying, the treatment in the field was to spit in the dirt, work up some mud, and apply it to the stings. It worked great.

Although I loved them all, my uncle was my favorite because he was a kid at heart and it was like hanging out with a buddy. Since he graduated from high school in 1947 and I was born in 1947, he used to say that 1947 was a great year for graduates but a poor year for babies. He was always pulling some prank on me and my cousins. The stars look amazing, he told us, if you put your coat over your head and peer up the sleeve at the heavens – and then he'd pour a glass of water down the sleeve. He was always sending us on missions to find a sky hook or green lantern oil, or some other such thing that didn't exist.

I remember his tale of the side-hill wampuses. He told us about them one night as we set up a tent in the woods to sleep out. They were a weird creature, he said, with one front leg and one hind leg on the same side of their body shorter than the other two. This was so they could walk on side hills and remain level. I guess I never thought to ask what happened if they decided to turn around and go in the opposite direction! Of course, we would also be visited by someone scratching on the side of the tent in the middle of the night – guess who?

I could go on and on about all of the things we would do, such as catch trout in a stream without a fishing pole, or drive down the road in his old Jeep and act like Forrest Gump (before there was a Forrest Gump), or act like some other lunatic while feeding the cows and doing chores. My uncle made us laugh. He filled us with enthusiasm. Even work was fun, and therefore more productive.

We all worked hard for long hours. Here are some of the tasks we either did, or observed so we could learn:

SEASONAL TASKS:

Bringing in firewood from the woods and stacking it in the cellar

Removing the horns from the young heifers

Shearing sheep

Hunting down wild dog packs who would kill the sheep

Mending fences using locust fence posts made from trees on the farm

Picking stone off the fields for various uses

Repairing building foundations with field stone

HARVESTING CROPS:

Cutting, raking, and baling hay

Bringing loaded hay wagons down steep approaches to the barn

Stacking hay and straw in the barn

Grinding corn for feed

Thrashing buckwheat

Combining oats and wheat

TENDING TO COWS AND MILKING:

Feeding the cows grain and hay

Feeding the calves with buckets of milk

Carrying milk from the main milking area to the milk house

Getting a two-quart can of fresh milk daily

Putting filled milk cans in the water cooler

Washing the milking machines

Helping the milk man load the full cans onto his truck while
exchanging stories

Treating cows for various afflictions such and mastitis,
pink eye, skin grubs

Cleaning the gutters behind the stanchioned cows

Fluffing up the straw in each cow's stanchion

Spreading manure

MAINTAINING EQUIPMENT:

Learning to weld

Tuning engines

Changing oil and lubricating

Repairing chains and belts

Repairing tires

Building hay wagons

RECREATION:

Carving your initials, date, and height on a tree in the woods

Playing ping pong, pool, or horseshoes daily

Building a dam for a swimming hole

Hunting deer, turkey, fox, and raccoon

Fishing and swimming in the farm pond

Fishing the trout streams

Going to town on a rainy day (Canisteo or Hornell)

Playing cards at a neighbor's home

Bowling

Making homemade ice cream

I learned so very much that I wouldn't have learned by living solely in the village. The Mennonite community in our area today often reminds me of those days. I believe so many people today are missing so much: the sweet smell of fresh-cut hay, the vigor from working on the farm, the social joys of getting together with neighbors, and the dependence upon God of your understanding for strength and faith and weather to sustain the crops. For those things I'm forever thankful.

– Jerry Kernahan, Penn Yan, N.Y.

CHAPTER 4

YOUR FARM'S
FINANCIAL PLAN

"The farmer is the only man
in our economy who buys everything at
retail, sells everything at wholesale,
and pays the freight both ways."

– John F. Kennedy

YOUR FARM'S FINANCIAL PLAN

T he life of a farm boy or farm girl seems idyllic: racing through pastures, riding the wagon, jumping in the mow. We remember the joy and tend to forget any troubles. Those days indeed may have been carefree for children, but mom and dad without a doubt had concerns aplenty – whether it was a storm coming as the clover lay in windrows, or a cow down with milk fever, or a horse in colic. And you can be sure they spent hours poring over the bank books, trying to figure how to make a go of it. To realize the dream of farming, one must be a good steward of the finances.

In this chapter we will take a look at some best practices in keeping the books, but let's start with a pair of reminiscences on rural childhood, beautiful in their fondness for simple joys, in their appreciation of working hard to get ahead, and in their admiration for the adults in their world who were able to make it all happen.

WE'RE OFF TO THE FAIR!

I can still hear Grandma Gibson singing: "Climb, climb up sunshine mountain, faces all aglow." She sang it as we climbed the slope in back of the barn whenever she came to visit our farm near Phillips Creek, and she'd play it on the piano after we returned.

I treasured her visits – and I treasured life on that farm. Our family was very close, and we all did our share of chores and hard work. As a child, there was plenty of exciting things to do to keep busy.

On Sundays, we'd go for a picnic in the woods. We had a favorite spot among a stand of pines. We would build a fire and cook hot dogs or whatever we had. We would venture out into the woods and hunt for the amazing "pink lady slippers." My sister Priscilla and I were always delighted by their beauty.

The highlight of the year on the farm was going to the Allegany County Fair. We each got to take our favorite cow or calf to show. The big excitement was getting to stay all night in the stall with your calf. The fair was a big deal for us, though there were plenty of things we were told to stay away from. Of course we did as we were told – ha!

You never get the farm life out of your blood. Taking care of livestock, digging in the dirt for your next meal, and enjoying the closeness of your family – it's simply the best!

– Barbara Fleishman, Belmont, N.Y.

A DAY AT THE RACETRACK WITH DAD

My father owned and trained harness racing horses, which meant he spent every summer at one of the tracks. He took my older brothers with him, which made me jealous because I wanted to go, too. Finally, when I turned 13, he decided I was old enough to go, so off we went to Vernon Downs. I was both thrilled and petrified. See, I had one small problem: I was afraid of horses.

Let's face it, I was small, and the horses were big. At 4 foot 10 inches, I was vertically challenged, to say the least. Still, I learned to do all the little things necessary to take care of the horses like cleaning stalls, rolling bandages and cleaning the harness. Although I was still scared, I started to get more comfortable being around these giant animals.

And more importantly, I got to spend time with my dad.

One day after coming off the track, Dad asked me to cool down our horse named Mountain Man. He was one of our bigger horses, and I could feel my fear starting to grow as I began walking him around. Much to my surprise, he acted calmly and just followed me around as he cooled off.

"Wow, I'm actually doing this," I thought to myself as I led the horse. "This isn't so bad after all."

I felt proud.

Then a gust of wind blew a paper bag in front of us. It startled Mountain Man, and he reared, taking me with him like a human yo-yo. I could hear my dad yelling as he ran toward us: "Don't let go of that horse!" In my terror, it seemed as if he cared more about the horse than about me.

Dad got the horse under control and turned it over to my brother. Then he turned and put his arms around me.

"I am so glad you didn't let go," he said as he hugged me tightly. "I was afraid that you were going to get trampled." We both breathed a huge sigh of relief and got back to the business at hand. After all, there was still work to do.

I got over my fear of horses that summer. And I spent precious time with my father. It was a summer that I'll always remember.

— Mary Ann Hoffman, Wellsville, N.Y.

THE FUNDS TO GET STARTED

When starting your farm, it is important to consider how you will finance it. Are you able to finance the farm yourself through a spouse or partner's off-farm income or your own accumulated saving and investments? Could you sell part of your existing property or other personal assets?

Ideally, it is best to avoid going into major debt when you are starting your farm. Yes, there are things that you will need to meet your goals, but if you bury yourself in debt right out of the gate, you're inviting unneeded stress and increasing the chance for failure. As the saying goes, "better to eat the elephant one bite at a time rather than trying to swallow it whole".

What about family and friends? Are you able to borrow from them? If that is an option, it could provide you with a reduced-interest loan, but you should always make sure to have everything in writing.

Taking out a second mortgage or home equity loan might be another avenue to consider for financing your farm. For many people, this is a frightening thought, but if it is your only source of capital and you truly want to start a small-farm enterprise, it could be worth consideration.

Another option would be to borrow from a local bank, credit union or other lenders that cater to farmers such as Farm Credit. If you think that might be an option for you, begin with lenders who have some experience in dealing with farms and other agricultural businesses. What about other small-business lenders that might be in your area? This includes micro enterprise funds. Economic and business development organizations in your region may be able to provide you with sources for a low-interest loan.

You should also consider grant opportunities on the local, state and federal levels. There are many grants targeted at assisting farmers in diversifying their enterprises while also keeping small farms profitable and even encouraging sustainable agriculture.

DEVELOPING A BUSINESS PLAN

Before starting a small farm, or opening any type of business, you should have a solid business plan. A business plan is essentially a road map to direct you on your path toward your goals and provide a reference point from which to return should you get off track or need to make adjustments. The basic components of your plan should include:

* Goals of your small farm
* Production plan (including materials and labor)
* Financial plan
* Marketing plan
* Follow-up plan

ESTIMATING YOUR SALES

It is imperative to realistically estimate your sales based on sound research. Keep in mind that the accuracy of your estimates will be directly correlated to your research. Little research equals poor estimates, whereas more thorough research will provide you with better estimates.

One of the most popular options for selling produce from a small farm is at a roadside stand. When estimating sales from such an operation, you must take into consideration that sales will typically vary not only throughout the day but also through the week and from one season to another. With our alpaca products, we usually see a spike starting in the fall lasting through the holidays. We need to ensure we have ample stock on hand to meet demand, with some products needing to be made or ordered the prior spring to ensure we have them in time.

If you are considering selling your produce to a retail outlet or food service venue, it is a good idea to conduct a survey of several within your area to determine the local demand and how it might vary by week and season. As you conduct market research, remember that you should also take into consideration a best and worst-case scenario. This will help you to recognize what you might reasonably expect in terms of profit.

THE IMPORTANCE OF DIVERSIFICATION

Any experienced financial planner will stress to you the importance of not placing all of your savings into a single investment vehicle but instead diversifying your investments to reduce your risk and increase your potential for return.

An investment portfolio that is well diversified will include a mix of high-risk, high-return investments along with some that are low-risk and low-return.

The same principle should also apply to farming. When you diversify your operations into more than a single product, you can reduce your chances of having a bad year financially while also evening out your returns over the long term. Ultimately, diversification can provide your small farm with more stability.

MAINTAINING YOUR EDGE

A primary reason for lack of success in any business, including farming, is not following up on your competitive position. While you may start out with an edge over the competition, if you do not stay on top of things you could find that the market has changed dramatically over a short period. To make sure that the products from your farm maintain an advantage over the competition, you must monitor regularly. One way to do this is to ask for feedback from your customers, either through surveys or by just talking to them.

PRODUCTION PLANS

You also need to make sure that you have a solid production plan in place for your small farm business.

This will assist you in organizing activities needed to produce the commodity, projecting your cash flow, and, if necessary, hiring labor. Production needs, of course, vary from one commodity to another. Remember to include all costs such as fertilizer, repair and fuel for machinery, irrigation expenses, pollination costs, marketing expenses, and anything else that might be associated with your goods.

YOUR FINANCIAL PLAN

A solid financial plan for your farm should include a projected income statement, cash flow analysis, and a source and use-of-funds analysis.

INCOME STATEMENT

A projected income statement for a farm can be completed much in the same way you would determine your personal budget, which is by figuring the amount of money coming in, subtracting normal living expenses, and then determining the amount of money that is left for discretionary spending. With a small farm, you will need to begin with your projected sales, subtract all expenses, and determine the amount that remains for profit.

Business costs typically include costs associated with production, such as labor, fertilizer and disease control.

These are variable costs. There are also fixed costs or those that do not change with production, such as interest on loans and depreciation. When completing a projected income statement, it can be helpful to complete one for best and worst-case situations.

CASH FLOW ANALYSIS

A cash flow analysis is somewhat different from an income statement in that it involves the amount of cash coming in and out instead of revenues and expenses. In some instances, you may receive the cash for a sale when you make that sale. In other instances, you may not receive the cash until later. The same is also true for making purchases; in some cases you must pay immediately and in others you pay after delivery of the item. There may be times when more money is coming in than your expenses call for, which is a great feeling. However, you need to be disciplined enough to save this income for times when sales are slower. A cash flow analysis will help to identify those times when you have more money flowing in than others so you can maximize the best times for your farm.

SOURCE AND USE-OF-FUNDS ANALYSIS

You will want to identify where you expect your funds to come from in order to budget properly for needed equipment and supplies. When we started, we had some available cash from a property sale that allowed us to purchase our starter herd and provide some basic updates to our barn.

We identified many things that we also would need to make our farm operations manageable, such as a trailer for shows, tractor, additional fencing, etc. We worked with a local bank to arrange a line of credit so that we could continue to move forward. This was critical as it provided breathing room to continue our operations without depleting our personal cash reserves. For the loan, we needed to identify each item and its anticipated cost, along with a cash-flow plan of how we were going to pay the loan back. Depending on your startup costs, you may be able to finance your operations from existing reserves, off-farm income, credit card, or line of credit. Just make sure that you identify those needed things in advance and have a plan as to how you are going to pay for them.

WHERE TO GET HELP

A great free resource that I encourage you to review is provided by SARE (Sustainable Agriculture Research & Education). It is called "A guide to developing a business plan for farms and rural businesses." It can be downloaded for free at **www.sare.org/Learning-Center/Books/Building-a-Sustainable-Business**, or you can pay for a hard copy. This useful guide provides all sorts of charts, spreadsheets, and analysis tools to help you in creating your farm's business, financial and marketing plans.

The U.S. Department of Agriculture also provides a wealth of valuable information on its website: **www.nal.usda.gov**.

Many agricultural software programs are available to help you keep track of your farm operations. Try an Internet search, or ask your accountant for a recommendation. QuickBooks is one of the more popular programs for accounting, and we use a program called "Alpaca Ease" or "Herd Ease." This has helped immensely with keeping track of our farm expenses and herd management, and it produces reports that our accountant can use to complete our schedule F for tax purposes. Go to: **www.alpacaease.com** for more information.

If you are more of a paper and pencil type person, then a general ledger from your local office supply store will do the job.

Remember "if it can't be measured, it can't be managed," and disorganization will lead to frustration. The key is to find something that will work and stick with it.

PROFESSIONAL ASSISTANCE

As you begin your farming venture, you may wish to consider adding three key members to your financial team.

★ A good financial advisor can help you with putting together a financial road map for your personal and business goals. The advisor may be able to help you with budgeting, suggesting other financing options, establishing a retirement plan, and making sure that you are properly insured.

★ If your farm is run as a true business, you may be eligible for certain tax benefits that are available to someone operating a small business. You will want to seek out a good accountant who has experience in working with farmers to ensure that you are setting up your books and records correctly. You will also want to become familiar with the rules of the game by reading the **IRS Publication 225, Farmer's Tax Guide**. A website link is provided in the index.

★ Although most small farms are operated as sole proprietorships, you may need the services of an attorney to set up the proper legal structure for your business such as an S Corporation or a Limited Liability Company (LLC).

The best way to find qualified people is simply to begin asking other farmers whom they use. You're likely to start hearing a few names repeated, and those may be good people to call. You may need to interview several before finding a match.

TO MARKET, TO MARKET ...

The world's best financial plan, however, won't help you if you don't have finances to manage. Nor will all your efforts toward production efficiencies and optimal use of the land do you any good if nobody knows about you or what you offer. You need to get the word out – as we will see in the next chapter.

CHAPTER 5

A MARKETING PLAN

"Sowing is not as difficult as reaping."

– Johann Wolfgang von Goethe

A MARKETING PLAN

E ssential to any successful farming operation is getting the word out about what you offer – your product, your service, or whatever makes you special and a notch or two above the competition. A solid marketing plan can help you develop a customer base upon which to earn a living.

An acronym that we use is ABM: Always Be Marketing. From business cards and T-shirts to your farm's website and social media outlets, you need to be promoting your farm and products.

In short, you want those who come to your farm or buy from it to feel its welcoming embrace and come back to do business with you – and that's a fundamental of marketing that the father in the following tale might have done well to heed:

OUTHOUSE BLUES

I think I was four or five when Dad picked up those darn chickens. The hens weren't so bad, but along came two mean roosters that became my worst enemies. When nature called, I leisurely made my way from our house to the outhouse to have a visit with the flies and the Sears catalog (toilet paper was scarce back then).

It was about 150 feet from house to outhouse, and on most days it was no problem to navigate – that is, until the roosters decided that I was fair game to attack at any moment on my journey. I would be walking along and seemingly out of nowhere they would descend on me, pecking my legs and arms and even knocking me down. Many were the times when I had to use the outhouse so badly that I thought I would burst. I would look out the window to make sure the roosters weren't around and take off running for the outhouse, slamming the door shut and peeking out through the holes to see if they had discovered me.

After I had suffered several weeks, Dad finally had enough of those mean old roosters and got rid of them. I was secretly hoping that they became someone's supper. Today I have a great appreciation for indoor plumbing.

– Don Nielsen, Bluff Point, N.Y.

WHAT A MARKETING PLAN SHOULD INCLUDE

When direct marketing your products, it is important to ensure you have a plan in place that will allow you to identify your competitive marketing advantage. A solid marketing plan should include:

- ★ Your product description
- ★ Your competitive advantage or USP (Unique Selling Proposition)
- ★ Your plans for promotion and marketing
- ★ Sales forecasts
- ★ A plan for maintaining your competitive position

YOUR PRODUCT DESCRIPTION

Defining what you sell is important and should be clear and concise; however, defining who you are and why your customers should buy from you is even more critical to your success. If someone were to ask you what you do for a living, could you give them a short, concise answer (sometimes called your "elevator pitch") to quickly identify what you do? Yes, you might be growing lettuce or raising pastured poultry as a product, but what a person is really buying is you. If you can't establish trust between you and your customers, it won't matter much what you produce: If they don't like you, they won't be buying from you.

YOUR COMPETITIVE ADVANTAGE OR USP (UNIQUE SELLING PROPOSITION)

What is unique about yourself or the product you are going to offer? Essentially, you need a hook of some kind that makes your product or service interesting and helps the consumer identify with you.

Our alpacas are unique in that there aren't very many of them and they are just plain cute. They generate a ton of questions and we have had to develop a standard list of answers to all of the basics, such as: "How many do you have? Can you ride them? Can you eat them? What is an alpaca, and what do you do with it?" Now if we could just get people to quit calling them llamas.

YOUR PLANS FOR PROMOTION AND MARKETING

Marketing costs are an important line item in your farm budget. You need to create a formal marketing plan so that you can allocate your budget prudently and get as much bang for your buck as possible. Some things are very cheap to implement, and others may need to wait until cash flow will support it. We stumbled along with some basic business cards and cheap farm banners until we could get a professional website built. We then began to advertise locally and put on an annual farm-day event for people to come out and see our animals. We then added a small farm store and rented out space in a local retail store that catered to small-business vendors.

We are getting our feet wet using social media, and we launched an online store. Find us at **www.cnlpacafarm.com**. There is no one thing that is the magic bullet for marketing your farm. It is the steady application of doing many things consistently over time that ultimately builds your recognition.

SALES FORECASTS

For new farmers, this is an area where you need to be brutally honest with yourself. It is easy to get caught up in the romance of how great you think things are going to be, only to be sadly disappointed by your actual results. We had grand dreams of having all kinds of money coming in from selling our alpaca offspring called "crias," stud fees from some of our males, and possibly some income from boarding other customers' animals. Now that we have been in business for almost five years, we have totally changed our game plan, moving away from the actual breeding end of the business and focusing more on making products from the elite fiber that these animals produce. Sales forecasts give you a goal to shoot for; however, you need to remain flexible and realistic with your business direction and its income potential.

A PLAN FOR MAINTAINING YOUR COMPETITIVE POSITION

One thing is for certain, if you become successful at what you do, you can bet there will be others waiting to move in and take away some of your competitive advantage.

That is why you need to always be looking for ways to keep your name and products in front of the right buyers. The Internet has opened the door wide, providing all sorts of ways for you to keep track of your competition, and you can bet they are doing the same to you. We routinely send out e-mails to our clients letting them know about events and products we have added to our website. We are also making plans for a quarterly newsletter. One of the easiest ways to stay one step ahead of your competition is by offering your customer a great buying experience. What can you do that will "wow" them to make them feel special? Maybe it's a surprise discount for loyal customers, or free shipping on an order, or a simple smile and heartfelt handshake to thank them for doing business with you. You need to figure out unique ways that will keep them coming back for more.

INTERNET AND SOCIAL MEDIA

Today, a large majority of consumers find information about you and your farm via the Internet. We have received orders from all over the country from people who have found us via an online search. Having a well-designed website for your farm is important if you want to be found. The cost of having a website built has come down significantly over the years, and you can even create a free website yourself at **www.weebly.com**.

The use of social media has exploded, and Facebook is the king. Having a social media presence for your farm can multiply the number of people who can find and recommend you.

People like to get to know you, so periodic updates about your farm can strengthen the relationship between you and your customer.

Like to write? Create a blog about your farm. The "Wordpress" (**www.wordpress.com**) or "Blogger" (**www. blogger.com**) platforms are free and could be a great way to connect with your customers.

NETWORKING OPPORTUNITIES

Numerous networking opportunities are available to get the word out about your farm and products. Consider these:

★ Joining your local Chamber of Commerce or state Farm Bureau can give you an instant connection to other businesses and farms.

★ Farmers' markets and Community Supported Agriculture (CSA) programs have the potential to help you reach consumers looking for your products.

★ Get to know your local newspapers. Connecting with a reporter for human interest or business stories can help to get the word out to local residents about your farm.

★ Join clubs, volunteer for committees, and support agricultural groups that can strengthen your bond to the community. The Future Farmers of America and 4-H are great organizations to support.

★ If your type of farm operation allows, offering an "open house" or "farm days" tour is a great way to invite the public to see what you have going on.

★ Having a booth at local and regional craft fairs and festivals can be a great way to gain exposure for your farm. Keep in mind that some of these are very competitive to get into and require advanced sign-up or committee approval before accepting your products.

★ Always keep business cards in your wallet or purse. These miniature billboards can have a wealth of information printed on them. You can also become a walking billboard, wearing hats, shirts, and other items imprinted with your logo. A great website for business cards and other promotional items is: **www.vistaprint.com**.

HOW MUCH CAN YOU AFFORD?

Keep in mind that some methods of marketing will naturally cost more than others. For instance, promoting your small farm through local radio and television outlets will be more expensive than other options, but on the other hand it will allow you to reach a larger market sector.

Always take into consideration how much you can reasonably afford to spend on marketing efforts. While marketing is essential for any business enterprise, it is important to factor in how much you can allocate from your resources, and you need to continue to review your marketing expenditures regularly.

Remember: ABM. Always Be Marketing.

CHAPTER 6

TROUBLESHOOTING PROBLEMS

"Farmers only worry during
the growing season, but townspeople
worry all the time."

— E.W. Howe

TROUBLESHOOTING PROBLEMS

Y ou can be as sure of one thing as you are of the sunrise and sunset or of the seasons changing: Things won't always go smoothly for you on your small farm. Take heart: You are not alone. In fact, through unity with other farmers, you will find the support you need to get back to business.

Here's a veterinarian's account of a most unusual problem and a rather unorthodox solution:

TURKEY CPR

As a veterinarian, I have fielded all sorts of emergency calls during my days, but once in a while you really get one that makes you laugh.

One Thanksgiving, I was on call and checked my answering machine. I heard a neighbor's panic-stricken voice. These people had moved to our rural area from a much more urban area in New Jersey. For some reason they had decided they wanted to farm. They probably had been watching "Green Acres" reruns. The comedy series, as you may remember, featured a city couple who bought a farm yet couldn't quite get away from the bright lights and Wall Street.

In our neighbors' case, the wife did seem genuinely interested in getting "back to the land." Her husband wasn't around very much, but he seemed friendly enough. He said he was a swimming pool salesman who traveled the South, and he carried a revolver on his hip. They were neighbors only a few years, then suddenly were gone. I had to wonder if they were in a witness protection program.

So back to that panicky phone message. It was something about a turkey. I couldn't decipher the rest. I had to leave, so I asked my wife to return the call. She could contact me on the radio if necessary. Besides, I knew my wife was more qualified to answer a turkey question, especially on Thanksgiving.

As I was driving down the road to treat a cow with milk fever, my wife called on the radio. Now what, I thought. She was laughing so hard that I had trouble understanding her.

Seems our neighbor's pet turkey had become ensnared in some electric fence and apparently had expired. Thus, the panicked message: What could she do?

When my wife returned the call, my neighbor boomed out the good news: "It's OK! I don't need the doc to come out. I bet you didn't know you can give a turkey CPR – mouth to beak!"

I don't think even I could have come up with that solution. Our neighbors' Thanksgiving was saved, through a little horse sense and a do-it-yourself attitude.

– John Roemmelt, DVM, Union Springs, N.Y.

The following writer tells of a common farm problem that he encountered as a youth – and one that fills him with dread to this day:

WHY I DON'T LIKE RATS

I hate rats. In fact, I fear rats more than any other animal found on the farm, and here is why. As a young man growing up on a beef farm in central New York, we fed corn silage. Corn silage is a cheap feed which is high in energy, and this feed will put pounds onto cattle, and that was the name of the game. Corn silage used to always be fermented in an upright silo. Our silo was 16 feet in diameter, and stood about 60 feet tall. By today's standards, it was not a big silo.

However, as a young boy of about 14, my job was to pitch the silage from the top of the silo down the chute, onto the floor beneath. I would have to pitch about 1,500 pounds of silage per day, and I did this every day without fail.

For some reason, we had a rat problem in the silo. Rats are nimble climbers which can easily reach the top of a silo, and there were rats in this huge pile of fermented feed. Each day, I would bang a shovel on the bottom door of the silo just to let the vermin know that I was about to begin my daily climb to the top. I would usually yell a few words up the dark cavernous chute to let them know I was on my way.

And then one day, as I was climbing up, a rat decided to head down. The rat went right into my coveralls, and since the legs of my coveralls were stuffed into my boots, the rat had no way out. I held onto the silo steps, screaming, as a large Norway squealed and wiggled about inside my coveralls. I was mortified, and to this day, do not know how I endured it.

In record time, I came down the silo ladder (at least 4o feet), and I don't remember even touching the rungs. Once at the bottom, I ripped open my outer garments, and the rat quickly exited my coveralls. I remember that long, nasty tail hitting my chin as it departed.

To this day, I do not get along with rats.

–John Kriese, Branchport, N.Y.

OVERCOMING OBSTACLES

As with any business enterprise, there are going to be times when you encounter problems with your farm. Small farms are commonly described as limited resource farms, but that does not mean there's a limit to the creative solutions you can come up with to solve potential problems. Your small farm can be profitable, but you will need to learn to be resourceful to overcome certain obstacles, including the following:

★ Limited purchasing power (no discount and small quantities)
★ Limited market availability (low volume)
★ Limited availability for custom field work (small fields)
★ Limited farm experience and knowledge (new to farming and agriculture)
★ Limited resources (money, equipment, land, etc.)

THE PROBLEM OF HIGH OVERHEAD

In some instances, a small farm can fail as a result of high overhead. This can include new equipment such as tractors, fencing, ornamentation, and even the construction of a new, large barn.

All of these elements can make it hard for a limited resource farm to overcome the related expenses.

If you make a purchase for the farm, it should fit into your overall business plan. If the farm is not able to support that purchase, you should not buy it. Better to take it slow and add things as time and capital will allow rather than to bury yourself in debt.

It is essential that you view your farm as a business and make sound business decisions regarding your operation. Given the limitations under which most small farms must operate, it is imperative that you be resourceful. Below are some ideas on how you can do this:

★ Utilize or repurpose existing materials and resources.
★ Utilize less-expensive fencing, covers, buildings, etc., when possible.
★ Purchase used equipment (a farm auction is a great place to pick up used equipment).
★ Purchase equipment that offers flexibility and can be used for more than one task.
★ Unless you can buy at a substantial discount, never buy more than you need.
★ Consider whether it is less expensive to outsource.
★ Consider whether it is cheaper to buy a product instead of produce it on your own.

Remember that while small farming can be highly rewarding personally, your goal, as in any other business, is to make a profit. So unless you are just doing it as a hobby, if you are not able to cover the variable costs of your farming enterprise, you should not raise it or grow it.

BENEFITING FROM A COOPERATIVE

In some instances, participating in a farm cooperative can help you in a variety of ways through the sharing of activities, goals, and the objectives of the members of the cooperative. Here are some of the benefits:

★ Group marketing strategies increase your visibility.

★ Group purchasing makes it possible to lower costs.

★ Diversity of goods produced can strengthen the small-farm community.

★ You can quickly develop a niche market.

★ Barter agreements and service contracts can establish reliable sources of farm labor.

★ You can explore newly identified markets.

★ You can reach a wider market base with untapped large producers.

★ Networking will help you to share experiences.

★ It's easier to develop and implement educational programs.

TIPS FOR A MORE SUCCESSFUL SMALL FARM

Working with a cooperative can maximize your opportunities and could help you to lower the costs associated with purchasing supplies, while also opening marketing avenues for your commodities. But it's just one of many things you can do to make your small farm more successful – the common denominator among them is getting to know people.

It only makes sense that the more people that know you the better the chance of connecting with them as potential customers or business allies. You need to make friends with your neighbors and fellow farmers. If you are new to small farming, do not hesitate to solicit advice and help. You might be surprised at how much you can learn from others. Be willing to participate in various agricultural organizations and associations in your area. They can prove to be great educational resources and provide you with information on other resources. Many of these organizations are listed in the resource section at the end of the book.

By establishing yourself as an earnest participant in a community of farmers, you can tap their collective wisdom and experiences – and forestall problems before you have to troubleshoot them.

CONCLUSION

TIME TO GO

"There seem to be
but three ways for a nation
to acquire wealth. The first is by war ...
the second by commerce ... the third by
agriculture, the only honest way."

— Benjamin Franklin

TIME TO GO

S oon we must part, dear reader, and get back to the business of running our small farms. In any farmer's life, there comes a time when you just have to go.

Witness the following story:

AT THE COW AUCTION

Many years back at a cow auction, I was waiting an exceptionally long time for the cow I was interested in buying to come up for bids. I'd had a few glasses of ice tea, and my bladder was full. Not wanting to miss my chance to bid, I held it as long as I could, and finally when the bidding was over, I hustled out to the back of the tent to relieve myself. I found what I thought was a quiet place.

As I took care of the matter at hand, I began to sing an old Hank Williams song that had been stuck in my head that day: "Goodbye Joe, me gotta go, me oh my oh ..."

On my second time through the verse, a woman who had exited the tent from a different direction happened upon me and let out a loud "oh my!" of her own before scurrying off the way she'd come. I was a little embarrassed but had to chuckle. I do hope she was able to recover.

– James McNinch Sr., Belmont, N.Y.

In this basic guide to small farming, I hope that I have helped to inspire you to launch your own enterprise by providing you with fundamental information on establishing goals, developing a marketing and preliminary financial plan, and understanding some of the problems inherent in small farming. If we can start out as newbie farmers and make a go of it, then so can you. My family has come a long way in our small farm journey and we still have much to learn. What we have learned though is that it takes teamwork and perseverance plus the willingness to keep trying new things in order to be successful.

Reviewing and revising your plans and your operating procedures regularly as well as taking advantage of the myriad resources that are available to the farming community, including the advice and guidance of experienced and established farmers, can help you to meet your goals and grow your farm. You'll find a wealth of such resources listed at the end of this book and also online at **www.farmwhispers.com**.

But before we go, I want to share with you a few more memories from people that you may feel that you already know. The following are two accounts of farm life that I find striking in their descriptive mood. They are, in short, what it's all about. Meet a farmer unable to sleep on a stormy night, worried about his chickens; and a veterinarian, alone in the barn on Christmas Eve. These are timeless stories that truly whisper to the heart:

FOR TUNA

I just had one of those epiphany moments. Actually it took nearly an hour, but the awakening may last the rest of my life. I'll see if I can capture it clearly in black and white.

Why was I restless? Why could I not sleep? Tired from a long weekend, I tossed and turned in my bed. Awakened by the passing cars, a whimpering daughter, thoughts of castrating newly born piglets, and eventually the pounding rain, I found myself in and out of bed for most of the young night's sleep.

I had always heard that farmers pray for rain. I do. I love to see the grass grow and the soil take on the softness of life. I love to see cows turn grass into many amazing things. A friend recently told me he wants to write about sustainable farming and the "cultural mandate" in Genesis 1:28. It all fits. God provides rain, and we tend and till His creation.

Yet this year, unlike no other, we had fallen victim to the rain as well. About three weeks ago we lost 160 chicks to rain. It appeared to be a catastrophic loss to our plan to raise enough chickens to pay the bills for the farm.

God has been taking me to the end of my safety line and asking me to let go. When am I going to fall off? Is there a way out of this tailspin? Why are we in Ohio? Why did you lead us into the desert to starve? It seems more like the curse of thorns and sweat in Genesis 3 than the mandate in Genesis 1.

Then tonight, it was pouring. I haven't heard it rain that hard for that long in quite a while. It crossed my mind to roll over and go to sleep – like I did that last time. "Surely, the chickens have been out long enough to have all learned how to find shelter under the cover of their pen."

I have had many such thoughts as I have learned to farm. Which ones do I pay attention to? For instance, "I don't think those cows will eat that brush pile comprised of yew bushes."

Or in another case, "One night with the door open to the coop won't hurt the chickens." In both cases, I suffered losses that caused damage to the ship.

This night would be different. I am learning. Can I really ignore those realities any longer? If I do not go and check on the chickens, we might suffer another loss. So, I grabbed a pair of shorts (turned out to be a swimsuit – good thing!), a shirt, my sandals, a rain coat, and headed out the door. I was wet within ten feet out the door. It was raining!

I hopped over the fence and into the pasture heading toward our older chickens and had to move about 25 birds to shelter. Why do they not go to cover? I have no idea.

Then, I turned to our youngest flock. It was a repeat of the scene from a few weeks ago. There were piles of wet chicks in the open end of the pen! Despite my prior experience, this was still unexpected. I went in under the pen on my hands and knees and started moving chicks. After finishing one pen of about 100 chicks and moving to another, I realized that this might not work. They were too cold and wet and were in need of greater care.

Turning toward the house to get the truck and some containers to carry the chicks back to the barn, my first thought was, "I can't do this by myself." After hopping back over the electric fence and running across the yard, I realized that I was just alone and wanted to be consoled. "Not another loss. Why?"

I hadn't seen this part of farming before we started. I didn't realize the darkness that might come. I didn't realize that I would feel the weight of responsibility for my family, our finances, and the weather! "Where are you, God? Why are you are doing this to me? You called me to this life. For what, to teach me that chickens are stupid?!"

Instead of getting Mel, I got containers and the truck. Back through the fields I drove to start loading chicks so I could put them under heat lamps in the barn. It wasn't until midway back to the barn that I realized the song that was playing on my truck CD player. A friend had made the CD for his mother. He labeled it "For Tuna," because that is her nickname. That label has so much more significance.

The song is titled, *"Come Awake"* by David Crowder. Here are the lyrics:

Are we left here on our own?
Can you feel when your last breath is gone?
Night is weighing heavy now
Be quiet and wait for a voice that will say
Come awake, from sleep arise
You were dead, become alive
Wake up, wake up, open your eyes
Climb from your grave into the light
Bring us back to life
You are not the only one who feels like the only one
Night soon will be lifted, friend
Just be quiet and wait for a voice that will say

Rise, rise, to life, to life
Shine
Light will shine
Love will rise
Light will shine, shine, shine, shine
He's shining on us now

Those chickens needed some serious heat from a serious light. The sun wasn't due to rise for six hours. Their light was in the barn, and it might not even be enough. But our light has risen! He is raising us up. He is the sustainer of all things, even agriculture. He is the restorer of life. He has broken the curse in Genesis 3 by wearing a crown of thorns and bleeding blood out of the sweat glands of his brow! He claims in Hebrews 13, "Never will I leave you. Never will I forsake you."

Is this true? I think so. Is it coincidence that "Come Awake" was playing in a dark hour? I think not. Does this mean that I won't go out tomorrow and find many dead chicks? No. But it does mean that I am not the only one who feels like the only one. He is shining on us now! Amen!

– Steve Montgomery, Lamppost Farm, Ohio

A CHRISTMAS EVE TO REMEMBER

I had been out all day treating sick cows on that Christmas Eve in 1984, and it was one of those cold and snowy nights in New York's Finger Lakes region. I was exhausted. I wanted nothing more than a relaxing evening, my veterinarian duties done. My wife greeted me at the door with a big smile and was busy preparing a special holiday meal.

I asked my daughter if she had done the evening barn chores yet. She had not, so I told her that on this special night, since I already had my coveralls on, I would do them for her.

Down I went to the barn that I had built with my own hands. I flicked on the lights. And there they were: There was Hopeless, the Hereford-Texas longhorn cross steer, destined to become an ox as great as Babe. Redford, our beef cow, and her calf Raja.

There was Sheeba, our daughter's first horse. We had gotten her from the vet clinic for free, and she was worth every penny. Junior, our daughter's 1,200-pound pride-and-joy Appaloosa gelding – nice horse, but sometimes one wondered if he was blind or just inclined to wander into things.

And you can't forget Dumb Delbert – I mean, how can you not love a horse with three nostrils? Once he had somehow managed to get his nose hooked on a water bucket.

Looking around, I saw our odd bunch of barn cats – some even with names. My good old truck-riding dog was absent, though; he had been smart enough to stay in the warm house to eat his supper.

I fed them all, topped off the water buckets, and sat down on a hay bale. As I sat there, all I could hear was the sound of animals chewing. I sat there for fifteen minutes just listening, reflecting.

So this is what it must have been like, I thought, in the stable on that first Christmas Eve so long ago.

–John Roemmelt, DVM, Union Springs, N.Y.

High time to get to work: The day will be over all too soon, if you're like most farmers I know. That's partly because there's so much work to do, indeed. But it has a lot to do with the way sheer joy makes time fly. Like the star struck bull in our parting tale, hard work can be its own reward.

ONE HAPPY BULL

We had just purchased a young bull with some good genetics that we wanted to introduce into our herd. It was still young, but showed great enthusiasm when out in the pasture next to the heifers.

The day finally came when it was time to turn this new bull out into pasture with the heifers to do what bulls naturally do. We had twenty or so females that needed to be bred, and we figured that being a new bull it would take him a few days to do his work.

We had a ton of chores to do, so we turned him out early in the morning and figured we would come and check on him later in the day. We were all exhausted at the end of the day but went to check on the bull to make sure all was well. We scanned the pasture and saw all the heifers – but no bull. Our first thought was that he had gotten out through a hole in the fence and was somewhere roaming the countryside. We spread out and did a quick search, but no bull was to be found.

It was getting dark and harder to see, but I thought I could make out a dark lump over in a corner of the pasture that we hadn't checked yet. We went over to investigate, and sure enough there was the bull, lying in a heap. We thought at first he was dead. His tongue was hanging out, and his eyes were glassy. But he was still breathing. We wondered if he'd suffered a heat stroke.

After further investigation, we realized he was simply exhausted. You see, that young bull managed to impregnate all of those heifers in one day.

Now that's one happy bull!

– Glenn Andersen, Penn Yan, N.Y.

RESOURCES

For new farmers and ranchers, a wealth of resources is available to help with securing land, planning, and ultimately making a success of their farming or ranching enterprises. Some states and counties even offer special incentives and aid to encourage new farmers. Such resources can be particularly helpful for those who do not have a farming background. The government-supported agencies below all provide educational and technical support and assistance to the farming community:

NATURAL RESOURCE CONSERVATION SERVICE
Postal Mail: Natural Resources Conservation Service
Attn: Public Affairs Division, P.O. Box 2890,
Washington, DC 20013
Street Address: Natural Resources Conservation Service, 14th and Independence Avenue SW, Washington, DC 20250
www.nrcs.usda.gov

FARM SERVICE AGENCY

U.S. Department of Agriculture
Farm Service Agency
Public Affairs Staff
1400 Independence Ave., S.W.
STOP 0506
Washington, DC 20250-0506
www.fsa.usda.gov

The FSA also maintains an office in each state. The FSA offers two financing programs for land purchase by socially disadvantaged and beginning farmers. The new Farm Bill offers the Land Contract Guarantee Program and the Direct Farm Ownership Loan Program. The Beginning Farmer and Rancher Land Contract Guarantee Program was established due to the fact that traditional methods for farm entry and farm succession are no longer viable. This is a pilot program that is available in Indiana, Oregon, North Dakota, Wisconsin, Iowa, and Pennsylvania.

ALABAMA SUSTAINABLE AGRICULTURE NETWORK

Operates a Farmer-to-Farmer Program that links experienced farmer-mentors with beginning growers. The program requires new farmers to pass on what they have learned to others.
Telephone 256-520-2400
www.asanonline.org/mentorprogram.htm

APPALACHIAN SUSTAINABLE AGRICULTURE PROJECT

Supports farmers as well as rural communities within the mountains of western North Carolina in addition to the southern Appalachians with mentoring, education, promotion and community.

828-236-1282

www.asapconnections.org

CAROLINA FARM STEWARDSHIP ASSOCIATION

Provides farm incubators, which are educational farms where new farmers can lease a section of land and obtain access to advanced knowledge and equipment without the need to purchase it. Business planning assistance is also available through the program.

www.carolinafarmstewards.org/projects.shtml

THE FOOD TO BANK ON

A program that connects new sustainable farms in northwest Washington with mentors, raining and market support while offering fresh, high-quality food to the area needy.

www.sconnect.org/foodfarming/

GEORGIA ORGANICS' FARMER MENTORING & MARKETING

A program that combines established farmers with new farmers to transfer knowledge and expertise in sustainable growing. The program offers in-depth workshops as well as training in marketing, production and financial planning.
www.georgiaorganics.org/about_us/programs_projects.php

IDAHO-RURAL ROOTS

Supports sustainable and organic agriculture and community-based food systems in the Inland Northwest. Offers a Cultivating Success™ Program that provides existing and beginning farmers with decision making tools, planning tools and the production support and skills necessary to establish a sustainable small farm.
www.cultivatingsuccess.org

THE INTERVALE CENTER

Located in Burlington, Vermont the center offers two programs for new farmers. The Intervale Farms Program leases land and facilities to small organic enterprises and provides technical support with other more experienced farmers. Another program, Success on Farms works one-on-one with state farmers to strengthen their businesses.
www.intervale.org

IOWA BEGINNING FARMER CENTER

Coordinates education programs and services for new and retiring farmers. The center also assists farm families in developing skills in financial planning and management, tax laws, legal issues, leaderships, technical production, health, sustainable agriculture and the environment.
www.extension.iastate.edu/bfc

LAND STEWARDSHIP PROJECT

Sponsors Farm Beginnings classes in Minnesota, Nebraska, Illinois, North Dakota and Wisconsin each fall. Participants will learn financial planning, goal setting, marketing and sustainable farming techniques from established farmers as well as other professionals. During the spring and summer, participants attend farm field days and work with specific farmers.
www.landstewardshipproject.org/farmbeg.html

MAINE ORGANIC FARMERS AND GARDENERS ASSOCIATION

Operates the Journeyperson Program, which provides hands-on, mentored training to assist farmers in finding land where they can develop their business. The association also offers a Farmer in-Residence Program at the 250-acre Common Ground Education Center in Unity, Maine.
www.mofga.org

MINNESOTA FOOD ASSOCIATION'S NEW IMMIGRANT FARMER PROJECT

Provides training for aspiring farmers and new immigrants in micro-farming, gardening, marketing, land use planning and production farming.

www.mnfoodassociation.org/newimmigrant.aspx

ND SMALL FARM INSTITUTE

Offers information and training for aspiring, beginning, and transitioning farmers. The Growing New Farmers program in Belchertown, Mass. provides resources and services for new farmers in the Northeast.

www.growingnewfarmers.org

www.smallfarms.cornell.edu

www.nebeginningfarmers.org

www.nesare.org

NEW ENTRY SUSTAINABLE FARMING PROJECT

Assists immigrants and others with agricultural backgrounds to begin commercial agricultural enterprises in Massachusetts.

www.nesfp.org

NATIONAL IMMIGRANT FARMING INITIATIVE

Operates multiple programs for new farmers. The New Farmer Development Project educates and supports immigrants in New York City with agricultural experience to help them become local producers and establish small farms in the region.

www.immigrantfarming.org/index.php?page=New_Farmer_Development_Project

SOUTHEASTERN MASSACHUSETTS AGRICULTURAL PARTNERSHIP

A nonprofit organization that assists local agricultural enterprises achieve economic success. This is accomplished by creating locally produced agricultural products as well as providing business education to local agricultural enterprises.

www.semaponline.org

BEGIN FARMING OHIO

Columbus, OH: Public-Private Collaborative

http://www.beginfarmingohio.org/

BEGINNING FARMERS

http://beginningfarmers.org/

CENTER FOR FARM FINANCIAL MANAGEMENT

1-800-234-1111

http://www.cffm.umn.edu

FARM CREDIT SYSTEM FOUNDATION
Washington, DC.
http://www.fcsfoundation.org

GROWING SMALL FARMS
http://www.ces.ncsu.edu/chatham/ag/SustAg/index.html

NEW ENGLAND SMALL FARM INSTITUTE
Belchertown, MA
http://www.smallfarm.org

LINKS FOR NEW FARMERS
Manoa: University of Hawaii.
http://www.ctahr.hawaii.edu/sustainag/NewFarmer/Links.
asp

OREGON SMALL FARMS
Corvallis: Oregon State University Extension Service.
http://smallfarms.oregonstate.edu

SMALL FARM CENTER
Davis: University of California
http://www.sfc.ucdavis.edu

SMALL FARM CONNECTION
Puyallup: Washington State University.
http://smallfarms.wsu.edu

WOMEN'S AGRICULTURAL NETWORK

Burlington: University of Vermont.
http://www.uvm.edu/wagn

Along with the agencies mentioned above, other information sources also can provide tremendous assistance as you begin your farming venture. There are associations that also offer educational programs along with their association meetings. These can help you learn more about farming and find practical solutions to common problems, provided by people with hands-on experience.

OTHER USEFUL WEBSITES

★ NCAT Sustainable Agriculture Project – super farm resource with great information for beginning farmers: https://attra.ncat.org/

★ Micro Eco Farming provides a wealth of information for those looking to start or improve their own small farm: http://www.microecofarming.com/

★ The Rodale institute provides a wealth of information for organics and farm related information: http://www.rodaleinstitute.org/home

★ Eliot Coleman has been farming for over 40 years and his website and books are a must for anyone looking to grow their own food: http://www.fourseasonfarm.com/

★ Looking for recipes, ideas and tips from a community of like minded folks? This is the place:
http://www.homegrown.org/

★ Wonderful content rich site with a little of everything to help the experienced farmer to the newbie just getting started: http://www.motherearthnews.com/

★ Use this website to find farmers' markets, family farms, and other sources of sustainably grown food in your area, where you can buy produce, grass-fed meats, and many other goodies:
http://www.localharvest.org/

★ The National Sustainable Agriculture Coalition (NSAC) is an alliance of grassroots organizations that advocates for federal policy reform to advance the sustainability of agriculture, food systems, natural resources, and rural communities: http://sustainableagriculture.net/

★ The emerging slow money movement that's working to improve the health of local food systems and the economy: http://www.slowmoney.org/

★ The sustainable agriculture research and education website provides great content for farmers linking to information you need in your area of the country:
http://www.sare.org/

★ The US Department of Agriculture site is packed with information and is a must for anyone looking or researching anything to do with agriculture:
http://www.usda.gov/

★ Eat Wild is a website that will direct you to local farms that produce safe, healthy, natural and nutritious grass-fed beef, lamb, goats, bison, poultry, pork, dairy and other wild edibles.
http://www.eatwild.com/

ALPACAS

To learn about alpacas and the benefits of starting your own alpaca farm visit:
Find out what the McNinch family is up to on our farm page:
www.cnlpacafarm.com

A great general resource about alpacas:
www.alpacainfo.com

Research farms and animals in your area:
www.openherd.com

MAGAZINES AND PUBLICATIONS

www.smallfarmtoday.com

www.smallfarmersjournal.com

www.motherearthnews.com

www.grit.com

www.stockmangrassfarmer.net

www.maryjanesfarm.org

www.acresusa.com

IRS TAX GUIDE

www.irs.gov/publications/p225/index.html

COMMUNITY SUPPORTED AGRICULTURE (CSA) LOCATIONS

www.localharvest.org/csa/

BOOKS

You Can Farm – Joel Salatin

Family Friendly Farming – Joel Salatin

Pastured Poultry Profits – Joel Salatin

Micro Eco Farming – Barbara Berst Adams

Garden Answers – Rodale Press

Garden Problem Solver – Rodale Press

The Backyard Homestead – Storey

Making Your Small Farm Profitable – Ron Macher

The Winter Harvest Handbook – Eliot Coleman

Four Season Harvest – Eliot Coleman

Raising Poultry on Pasture: Ten years of success – APPA
Greenhouse Gardener's Companion – Shane Smith and
Marjorie C. Leggitt
Marketing Farm Products – Ellie Winslow
Alpacas: Synthesis of a Miracle – Michael Safley
The Complete Alpaca Book – Eric Hoffman

FARM AND GARDEN SUPPLIES:

Tractor Supply Company
www.tractorsupply.com

Growers Supply
www.growerssupply.com

Premier One Supplies
www.premier1supplies.com

PBS Animal Health
www.pbsanimalhealth.com
Light Livestock Equipment and Supply
www.lightlivestockequipment.com

SEED COMPANIES:
High Mowing Seeds
www.highmowingseeds.com

Sow True Seeds
www.sowtrueseed.com

Burpee
www.burpee.com

MARKETING AND PROMOTIONAL PRODUCTS:
www.vistaprint.com

CREDIT CARD PAYMENT PROCESSING:
https://squareup.com
http://payments.intuit.com

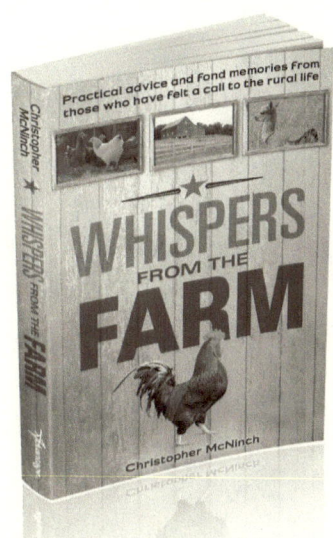

How can you use this book?

MOTIVATE

EDUCATE

THANK

INSPIRE

PROMOTE

CONNECT

Why have a custom version of *Whispers From The Farm*

- Build personal bonds with customers, prospects, employees, donors, and key constituencies
- Develop a long-lasting reminder of your event, milestone, or celebration
- Provide a keepsake that inspires change in behavior and change in lives
- Deliver the ultimate "thank you" gift that remains on coffee tables and bookshelves
- Generate the "wow" factor

Books are thoughtful gifts that provide a genuine sentiment that other promotional items cannot express. They promote employee discussions and interaction, reinforce an event's meaning or location, and they make a lasting impression. Use your book to say "Thank You" and show people that you care.

Whispers From The Farm is available in bulk quantities and in customized versions at special discounts for corporate, institutional, and educational purposes. To learn more please contact our Special Sales team at:

1.866.775.1696 • sales@advantageww.com • www.AdvantageSpecialSales.com

www.ingramcontent.com/pod-product-compliance
Lightning Source LLC
Chambersburg PA
CBHW050447290526
45786CB00006B/2191